MW00612662

The Foundations of
Newtonian Scholarship

Back row (right to left): A.R. Hall, J.B. Brackenridge, A.E. Shapiro, J.G. Fauvel and M. Nauenberg.
Front row (right to left): D.T. Whiteside, I.B. Cohen, M. Boas Hall, A.H. Cook, P. Harman, R.H. Dalitz and G.E. Smith.
(*Photo by David Fowler*).

The Foundations of
Newtonian Scholarship

Editors

Richard H. Dalitz
University of Oxford, UK

Michael Nauenberg
University of California, Santa Cruz

World Scientific
Singapore • New Jersey • London • Hong Kong

Published by

World Scientific Publishing Co. Pte. Ltd.

P O Box 128, Farrer Road, Singapore 912805

USA office: Suite 1B, 1060 Main Street, River Edge, NJ 07661

UK office: 57 Shelton Street, Covent Garden, London WC2H 9HE

Library of Congress Cataloging-in-Publication Data

The foundations of Newtonian scholarship / editors, Richard H. Dalitz, Michael Nauenberg.
p. cm.
Includes bibliographical references.
ISBN 9810239203 (alk. paper)
1. Newton, Isaac, Sir, 1642–1727--Congresses. 2. Physics--History--Congresses. I. Dalitz, R. H. (Richard Henry), 1925– II. Nauenberg, Michael.

Q16 .N55 F68 2000
530'.092--dc21 99-088303

British Library Cataloguing-in-Publication Data

A catalogue record for this book is available from the British Library.

Copyright © 2000 by World Scientific Publishing Co. Pte. Ltd.

All rights reserved. This book, or parts thereof, may not be reproduced in any form or by any means, electronic or mechanical, including photocopying, recording or any information storage and retrieval system now known or to be invented, without written permission from the Publisher.

For photocopying of material in this volume, please pay a copying fee through the Copyright Clearance Center, Inc., 222 Rosewood Drive, Danvers, MA 01923, USA. In this case permission to photocopy is not required from the publisher.

Printed in Singapore by Uto-Print

Mary and Rupert Hall talking with Tom Whiteside (seated) during a break in the morning talks.

Professor D. Fowler (left) with the editors M. Nauenberg and R. Dalitz (right).

Contributors' Biographies

J. Bruce Brackenridge is the Alice G. Chapman professor emeritus of history and physics at Lawrence University. His book *The Key to Newton's Dynamics: The Kepler Problem and the Principia* was published in 1995.

I. Bernard Cohen is the Victor S. Thomas professor (emeritus) of the history of science, Harvard University. Among his most recent publications are *Science and the Founding Fathers: Science in the Political Thought of Franklin, Jefferson, Adams, and Madison*, and a new translation of Newton's *Principia* together with a *Guide to the Principia.*

Richard Dalitz is an emeritus Royal Society research professor in the University of Oxford. He has published several articles on the history of modern particle physics; present research is in high-energy particle physics, in terms of quarks, leptons, bosons and gluons.

John Fauvel teaches history of mathematics at the Open University, UK, and is a former president of the British Society for the History of Mathematics. Among the books he has written for and co-edited are *The History of Mathematics: A Reader, Let Newton Be!, Mobius and His Band*, and *Oxford Figures: 800 Years of the Mathematical Sciences.*

A. Rupert Hall has taught the history of science at Cambridge and at several US universities. In the course of half a century's study of Newton's writings, he has been the senior editor of three volumes of Newton's correspondence and published numerous books and papers, including a recent biography of Newton.

Michael Nauenberg is a professor emeritus of physics at the University of California at Santa Cruz. He has written several articles on the works of Hooke, Newton and Huygens, and reviews of books on Newton's *Principia.* He recently collaborated with J. Bruce Brackenridge on an article on *Curvature in Newton's Dynamics* to appear in the *Cambridge Companion for Newton* (edited by I. B. Cohen and G. Smith).

Alan E. Shapiro is professor of the history of science and technology at the University of Minnesota. He is the editor of *The Optical Papers of Isaac Newton* and the author of *Fits Passions and Paroxysms: Physics, Method and Chemistry, and Newton's Theories of Colored Bodies and Fits of Easy Reflection.*

George E. Smith is a philosopher of science at Tufts University and a practicing engineer. His interests range from evidence in advanced sciences, engineering and medicine, and on the transition from speculation about the microphysical realm to systematic theory development during the course of late 19th and early 20th century physics and chemistry.

D. T. Whiteside is a professor at Cambridge University where he has been since 1956. He is the editor of eight volumes of *The Mathematical Papers of Isaac Newton* and of many seminal papers which have greatly elucidated Newton's work.

Curtis Wilson was a professor in the Department of History at the University of California, San Diego (1968–1973). He was the first recipient of the LeRoy E. Doggett Prize for writing in the history of astronomy, American Astronomical Society (1998). Among his publications are: *William Heytesbury: Medieval Logic and the Rise of Mathematical Physics* (Wisconsin, 1956), and chapters in *Planetary Astronomy from the Renaissance to the Rise of Astrophysics* (edited by Rene Taton and C. Wilson).

Preface

The initial conception of a *Symposium on the Foundations of Newtonian Scholarship* presented in this book, which was held in the Wellcome Hall of the Royal Society on 20 March 1997, stemmed from the realization that those interested in the life and work of Isaac Newton are living in a transition period. The eminent Newtonian scholars who have made accessible to us in a most remarkable way the vast mass of Newton's mathematical and scientific papers and correspondence, are still with us, mostly active and available for consultation, although in retirement or about to become so. Much new information about Newton's works has become readily available only in the last few decades and this has already led to significant recent developments in our understanding of Newton's *Principia* and of its origins and its proper interpretation on questions which had been obscure for several centuries. One editor (MN) and Professor Bruce Brackenridge, a member of the symposium organizing committee, have been fully aware of this situation for some time and have benefited from it, as this book attests. It was their desire to make this situation evident to scientists and mathematicians as well as to historians of science, young or established, which has led to this symposium. We hope that knowledge of these newly available sources may attract some of them to this research and that their work may lead to a deeper understanding of Newton's mathematical and scientific work and those of his contemporaries. This was a fundamental purpose of our symposium and its timing was motivated by the need for establishing contacts between new researchers thus attracted into this field and the eminent scholars who in the past have gained such a great familiarity with the development of Newton's scientific and mathematical style and abilities, and the lines of thought used by mathematicians in the late seventeenth century. This volume containing the proceedings of the symposium is also a tribute to their enormous accomplishments.

In 1985, the organizers of the Newton Tercentenary Conference, held at Cambridge University on 29 June–4 July 1987, proposed that Professor S. Chandrasekhar, the great astrophysicist, should give an opening address with the title "The Principia Today". Chandrasekhar agreed to do so, if this new endeavor turned out to be possible for him. At this conference, Chandrasekhar's talk on the *Principia* was quite brief, being only a preliminary report. He said that he had decided to select a number of the propositions from the *Principia*, to construct proofs of them using the modern methods he was a master of, and then to compare his proofs with those given by Newton. He reported only that he had been left, again and again, in "sheer wonder at the elegance, the careful arrangement, the imperial style, the incredible originality and above all the astonishing lightness of Newton's proofs." These proofs took for granted the knowledge and experience of a body of geometrical and other mathematical relationships, which Newton could draw upon when needed, but which Chandrasekhar, as well as most modern readers of the *Principia*, did not readily have at hand as part of his intuition. Newton's proofs were economical and highly original, devised by Newton for his immediate purpose. Chandrasekhar found Newton's treatments most admirable, and it was not long before he felt it urgent to make his conclusions widely known and to have other physicists and mathematicians join him in his admiration of Newton. Since Chandrasekhar had been a colleague of the second editor (RHD) at the University of Chicago, it was natural for RHD to ask him to visit Oxford University and give a series of lectures about Newton and his *Principia*. He gave 10 lectures (all of which were well-attended by Oxford standards) in the Trinity term of 1991; it was an occasion not to be missed! At the end of his visit, the Oxford University Press sought to print his lectures in book form, which he agreed to do, after he had expanded his lectures somewhat and had had further experience with the *Principia*. His book was published in 1995 under the title "*NEWTON'S PRINCIPIA for the Common Reader.*"

This book aroused much fresh interest about Newton, partly because Chandrasekhar laid out his treatments of the mathematical problems using modern methods well-known today, before comparing them with those of Newton. The *Principia* had always been considered a difficult book to read, suitable only for specialists and inaccessible even to other mathematicians and physicists, whereas Chandrasekhar's recasting of the propositions, lemmas and corollaries in modern terms made them readily intelligible and more readable. It has become known that, unfortunately, there are some serious errors in Chandrasekhar's book and that it is sometimes misleading, for he was not well-versed with the practice of mathematics in Newton's time, nor with the works of others beyond Newton at that time. Also, as he explicitly stated, he had made no use of the Foundation Works (see below), on which the papers of this symposium mostly depend. It has therefore been a controversial book, but a book which all those seriously interested in Newton and his *Principia* must read. Its great merit is that it has aroused much interest in this subject. Indeed, it has become a "best-seller," by the standards of academic books.

It is the custom for the Royal Society to require that a conference funded by its History of Science sub-committee should have at least one of its Fellows on its organizing committee. Given what has been recounted in the last paragraph above, it seemed natural that the second editor (RHD) should be the representative of the Royal Society for this symposium.

It is appropriate to add a few remarks concerning the publication of the Foundation Works:

The correspondence of Isaac Newton

This project was approved by the Council of the Royal Society in 1936, an editor was appointed in 1939, and work began in July of that year. But little progress was possible during World War II, and the pre-war editor died in 1946. A fresh start was made in 1947 when a Newton Letters Committee was set up to oversee the work; its chairman was Professor E.N. Andrade F.R.S., until his death on 6 June 1971. The project was given regular funding, from a grant obtained by the Royal Society for this purpose.

The editor for the first three volumes was Professor Herbert W. Turnbull, F.R.S., who died on 4 May 1961, just before Vol. 3 came out from the press. His assistant, Dr. Joseph F. Scott, edited Vol. 4 and continued as editor for Vol. 5 until his death in August 1971. Professor A. Rupert Hall and Dr. Laura Tilling were then appointed joint editors for the last three volumes, 5 to 7. The overall task was very considerable indeed. It had been decided to publish all of the letters which could be found, whether to or from Newton. According to Turnbull, Newton did not normally keep drafts of the letters he wrote during his Cambridge period, and it was necessary to search for his letters among those the recipients may have retained. According to Hall and Tilling, Newton often made many drafts during his London period; when the final letter was available, their comparison with these drafts showed that the last draft and the final letter were essentially the same. It was also decided to include letters between two other persons, where the contents (in part, if not in full) had bearing on Newton's work. Letters in Latin were to be published as such, with an English translation following each, together with endnotes concerning them. Further, notes jotted down by the sender and/or the recipient in preparing a letter and/or analyzing a letter received, were to be added, as appropriate. It was also necessary for the editors to understand correctly the matters referred to in all these letters, taking into acccount other developments in mathematics and astronomy going on at the same time; the editors of the last three volumes give thanks to Professor D.T. Whiteside for his advice about these matters, and for his good sense and scholarship in this respect. To emphasize the magnitude of the task, we give a short table to provide a quantitative summary of the outcome, over a period of almost 35 years of work.

Table 1 The correspondence of Isaac Newton.

Volume	1	2	3	4	5	6	7
Years	1661–1675	1676–1687	1688-1694	1694–1709	1709–1713	1713–1718	1718-1727
Publication	1959	1960	1961	1967	1975	1976	1981
pp.*	(**38**,468)	(**13**,552)	(**18**,445)	(**32**,577)	(**51**,439)	(**38**,499)	(**45**,522)
Editors	H.W. Turnbull			J.F. Scott	A.R. Hall & L. Tilling		

*In (**a**, b), **a** gives the prefatory pages, b gives the pages of text.

The mathematical papers of Isaac Newton

The publication of these papers followed a different pattern, rather unplanned. Remarkably, Professor D.T. Whiteside became the sole editor of the complete series of eight volumes from about 1960 to 1981, when the last volume came off the press. He recorded the supportive assistance he received from Dr. Michael Hoskin and Mr. Adolf Prag, the former mostly in the early years and the latter mostly in the later years. The funding of this project was irregular. Whiteside first came into contact with Sir Harold Hartley in 1959, in assisting him to put together and publish a book entitled "*The Origins and Founders of the Royal Society*" (Cambridge University Press, 1960) to mark the tercentenary of the foundation of the Royal Society in 1660. Hartley had always had a great interest in the history of science. When he saw what research Whiteside was doing for his Ph.D. thesis, he made it his business to keep Whiteside funded in the crucial period after completion of his dissertation — his research studentship ended in the early summer of 1959, and approval of his thesis for the Ph.D. degree by the Board of Research Studies of Cambridge University took place at their meeting in December 1960 — encouraging him and making it financially possible for him to continue his research, far beyond his dissertation, until it became the critical editing and the publishing of all of the mathematical papers of Isaac Newton. In the earlier part of his life, say up to 1960, Hartley had an unusually broad range of official and semi-official contacts and connections in Britain. With his unrivalled knowledge of the ropes, he was able to approach fruitfully all possible sources for the support of Whiteside and his project. By the time Hartley died (in 1971), Whiteside had become established and his funding had become regularly available, although from a number of independent sources and at a modest level. It is not surprising that, in our symposium, Whiteside's thoughts went back to those crucial early years, when he had been encouraged by, and supported through, his patron Hartley.

Whiteside's task was immense, for the mathematical papers were incomplete and unordered. He had to comprehend, date and correlate the various drafts and fragments of Newton's work as he moved towards the *Principia.* His ability to carry so many threads in his mind and to connect them correctly when needed is clearly most unusual. There are very few who could have done this work in such a reliable and thorough way. There are still obscurities about the degree of dependence between different papers to be understood and doubtless many pages were missing; what is remarkable is how much of Newton's calculations has survived and how Whiteside has been able to pull it all together and relate it with work going on by other mathematicians and scientists in Newton's time.

Most of Newton's papers are in Latin, and this original version is printed on the left pages of Whiteside's editions, together with an English translation on the right pages, although sometimes this translation is missing. The footnotes are very considerable, scholarly, and informative. Marvellously helpful is the Analytic Table of Contents at the front of each volume, typically being about 35 pages for about 500 pages of text. They give a precis from page to page through the text, giving enough information for the reader to readily find what he needs.

To illustrate the magnitude of Whiteside's achievement, the following summary table gives a quantitative measure of the final result, over a period of about 20 years.

Table 2 The Mathematical Papers of Isaac Newton (edited by D.T. Whiteside).

Volume	1	2	3	4	5	6	7	8
Years	1664–1666	1667–1670	1670–1673	1674–1684	1683–1684	1684–1691	1691–1695	1697–1722
Publication	1967	1968	1969	1971	1972	1974	1976	1981
pp.*	(**46**,590)	(**22**,520)	(**38**,576)	(**32**,678)	(**22**,627)	(**34**,614)	(**47**,706)	(**55**,704)

*In (**a**, b), **a** gives the prefatory pages, b gives the pages of text.

Other Foundation Works In This Century

I. Bernard Cohen, *Introduction to Newton's Principia* (Cambridge, Cambridge University Press, 1971).

Isaac Newton's *Philosophiae Naturalis Principia Mathematica* (1726 edition, with variant readings (eds.) I.B. Cohen & A. Koyre) two volumes (Cambridge, Cambridge University Press, 1972).

These three volumes go together. The latter two give a clear printing of the Latin text of the 1726 *Principia*, with brief footnotes (typically ten lines) in English. Pages 1 to 547 are in Vol. 1 and 548 to 771 in Vol. 2, together with ten Appendices, running from p. 775 to p. 916 in the latter. The variant readings display the differences in text between the 1726 edition of the *Principia* and the 1687 and 1713 editions, all of which which were directly supervised by Newton. These readings are the subject of the footnotes in Vols. 1 and 2 which contain also corrections and additions made by Newton into his personal copies of the first and second editions. A Guide is given on pp. ix to xl of Vol. 1. The MS for the two (1972) volumes was completed long before Koyre's death in 1964. The 1971 volume is exhaustive in its detailed documentation of the early manuscripts which led to the *Principia*, including discussion of those manuscripts with proposed revisions which Newton did not implement.

Isaac Newton, *Mathematical Principles of Natural Philosophy*, translated from Latin to English by I. Bernard Cohen and Anne Whitman with the assistance of Julia Budenz, with a Guide to the *Principia* by I. Bernard Cohen (Berkeley, Los Angeles, London, University of California Press, 1999).

The Optical Papers of Isaac Newton, Vol. 1 (*The Optical Lectures*, 1670–1672), edited by Alan E. Shapiro (Cambridge University Press, 1984); Vol. 2, in preparation. The first volume contains two versions of *The Optical Lectures* which Newton delivered at the University of Cambridge. One of these versions is the *Optica*, which Newton deposited in the University Library for public use, while the other version, the *Lectiones Opticae*, was kept by him and eventually became part of the Portsmouth collection. Both versions are reproduced in its original Latin text on the left page together with an English translation on the right page with extensive explanatory footnotes.

Isaac Newton, *Opticks Or A Treatise of the Reflections, Refractions, Inflections & Colours of Light*, based on the fourth edition London, 1704 (Dover publications, 1952). With a Foreword by Albert Einstein, an Introduction by Sir Edmund Whittaker, a Preface by I. Bernard Cohen and an Analytical Table of Contents prepared by Duane H.D. Roller.

John W. Herivel, *The Background to Newton's Principia: A Study of Newton's Dynamical Researches in the years 1664–1684* (Oxford, Clarendon Press, 1965).

A. Rupert Hall and Marie Boas Hall, *Unpublished Scientific Papers of Isaac Newton* (Cambridge University Press, 1962).

The original program for the symposium included a concluding talk to be given in the evening by Richard Samuel Westfall, the author of an excellent biography of Newton, "*Never at Rest.*" Unexpectedly, he died on 21 August 1996, a serious loss to us all, and unfortunate for the symposium. Professor Curtis Wilson agreed to give the concluding talk, but was prevented from attending the symposium by an illness at the last minute. Fortunately, he provided us with the manuscript for his talk which was delivered by D.T. Whiteside at the appointed time, with Marie Boas Hall as Chair. Both Rupert Hall and Whiteside were present, each speaking at, and contributing to, the symposium and its discussions. The other speakers gave talks illuminating various aspects of Newton's work, in the light of the Foundation Works.

The symposium was supported by grants from the Royal Society, through its History of Science sub-committee, from the National Science Foundation (U.S.A.) and from the History of Science Society (U.S.A.), and we thank them all for making this *Symposium on the Foundations of Newtonian Scholarship* possible. We would like to thank also Susan Johnson Cohen for taking most of the photographs of the speakers, which have been placed for each speaker on the page opposite the beginning of his lecture, and Professor David Fowler for the excellent group photograph which is placed on the page opposite this Preface. Above all, we are immensely grateful to the Royal Society for their permission to reproduce their fine portrait of Isaac Newton made by J. Vanderbank in 1726. We are very glad that it was possible for us to hold the symposium in the present building of the Royal Society, because of Newton's early and most important association with the Society, being its President from 1703 when Robert Hooke died, until the the end of his life in 1727. Today this association is made apparent by the five portraits, one of which is on the cover of this book, and one bust of Newton to be found in the Royal Society building.

Michael Nauenberg,
Physics Department,
University of California,
Santa Cruz CA 95064
U.S.A

Richard H. Dalitz,
Theoretical Physics Department,
Oxford University,
1 Keble Rd, Oxford OX1 3NP,
U.K.

Contents

Introduction

During the last half of the twentieth century, there have been outstanding advances in Newtonian scholarship because we have had universal access to major portions of Newton's papers and manuscripts that were previously available only to a few scholars. Among the major sources that have been published are the variorum edition of the *Principia* (1971) and the editions of Newton's correspondence (1959–1977), mathematical papers (1967–1981), and optical papers (1984).[1] Several recent contributions to the understanding of Newton's scientific work, based in part upon these sources, were presented by scholars at the symposium held in 1997 at the Royal Society in London. This book contains these contributions.

In the opening address, the distinguished Newtonian scholar I. Bernard Cohen reflected upon the foundations of Newtonian scholarship in a lecture entitled "Newton in Historical Perspective." He began with the following observation:

> "It was not until Rupert and Marie Hall published their volume entitled *Unpublished Scientific Papers of Isaac Newton* in 1962, that the scholarly world at large, the world of historians of science and of scientists and mathematicians interested in historical questions, became aware of some of the extraordinary new insights that were to be found by examining the unpublished Newton materials."

Most of these sources are now available in the seven-volume edition of *The Correspondence of Isaac Newton*, for which Rupert Hall was the major editor, and the monumental eight-volume edition of *The Mathematical Papers of Isaac Newton*, which was edited by D.T. Whiteside. In the latter edition, Whiteside not only provided a translation into English from Newton's original Latin text, but he also added scholarly annotations containing historical commentaries and mathematical clarifications. Both Rupert Hall and D.T. Whiteside participated as commentators at the Royal Society Symposium.

It is well-known that in the *Principia* and the *Opticks*, Newton established the foundations of dynamics and some elements of optics. He also solved and attempted to solve a number of very difficult problems in these two fields. Some of these problems remained outstanding during the centuries following Newton's death and became the major subject of scientific studies. In addition to his theoretical work, Newton also carried out remarkable but little known experiments on the diffraction of light and on the resistance of fluids; these experiments remained unmatched in accuracy for centuries, and stimulated further investigations. The lectures in this book contain a discussion of some of these problems, as well as the resolution of well-known puzzles related to the early evolution of the *Principia.* Examination of Newton's papers and correspondence with regard to his approach towards the solution of specific physical and mathematical problems, as well as to his early studies of dynamics and optics in general, provides an understanding of how the structure of his great masterpieces, the *Principia* and the *Opticks*, eventually emerged. One of the important new insights is an appreciation of the extent to which the organization of Newton's two books was guided by the major physical problems that he studied.

In his lecture "Newton's Experimental Investigation of Diffraction for the Opticks," Alan Shapiro described Newton's experiments on the diffraction of light from "slender obstacles", and his early attempts to explain the observations by his corpuscular theory of light. A hundred years after the first publication of the *Opticks*, Thomas Young made the following remark:[2]

> "The optical observations of Newton are yet unrivalled; and excepting some casual inaccuracies they only rise in our estimation as we compare them with later attempts to improve on them."

Some of Newton's unpublished optical papers[3] reveal the increasing care with which Newton carried out measurements of the position of diffraction fringes which led him to revise his optical models for diffraction. The remarkable accuracy of these measurements, which will be demonstrated here by a direct comparison of Newton's data with the modern theory of diffraction due to Fresnel,[4] led Newton to doubt his corpuscular models, and to resort ultimately to a set of Queries about the nature of light in the final section of his *Opticks.* In his review at the end of the symposium, Rupert Hall made the following comment:

> "Taken into the private laboratory, as it were, under Alan's guidance one meets a Newton who seems a good deal less dogmatic and confident than was the author of the *Opticks*, a book in which he had to face the public. This still uncertain Newton, seeking his way through a problem, who was exploring rather than legislating, choosing between multiple possibilities both mathematical and experimental that presented themselves to him, had to foresee and test his path leading towards the results he hoped to achieve."

In his lecture, "Fluid Resistance: Why Did Newton Change His Mind?" George Smith discussed the evolution of Newton's theoretical ideas and experiments on the motion of bodies in fluids like air and water. In 1926, a fellow of the Royal Society, R.G. Lunnon, repeated Newton's experiments on the resistance of fluids and made the following observation:[5]

> Newton's results are of much more than historical value, for two reasons. They are obtained from careful experiments which have never been repeated, and they contain a special value of the resistance constant which is a remarkably good one, well within the range of modern determinations.

In the second edition of the *Principia*, Newton made radical changes to the presentation given in the first edition, but he did not reveal the reason why he undertook these changes. As Smith shows, Newton undertook new experimental investigations on the resistance of fluids because of criticisms from his protege, Fatio de Duillier, but "sorting out what went on between the first and second editions of Book II would have been hopeless without Newton's correspondence ..." Comparing Newton's measurements of the resistance of fluids with modern results shows that they were remarkably accurate, but again, as in the case of the diffraction of light, Newton was unable to develop an adequate theory for his observation. Indeed in 1752, d'Alembert proved the paradoxical result that a Newtonian inviscid fluid does not provide any resistance at all, while an adequate explanation for the resistance of fluids was not given until 1904 by L. Prandtl. As Smith pointed out, the "resistance forces result from intertwined inertial and viscous actions ... a full account of this intertwining remains one of the unsolved problems of physics."

In his lecture, "Newton's Dynamics: The Diagram as a Diagnostic Device," Bruce Brackenridge takes up the role of curvature in Newton's early dynamics[6] as revealed in Newton's diagrams. In addition to the analysis of the diagram for Lemma 11 of Book I, the curvature lemma, particular attention is paid to the the recent analysis by Michael Nauenberg of a diagram in the 1679 correspondence between Newton and Hooke.[7] It may well be asked as to what can possibly be new to discuss in a topic as well studied as Newton's dynamics. As recently as 1991, Whiteside responded to that question as follows:[8]

> "Surely there can be nothing profoundly new to be said about its (the *Principia's*) progress from first conception as an inchoate idea in it's author's mind to the maturity of its first publication in 1687? No and yes. Anyone not of the fraternity, however, would surely be surprised to

> see how much Newton scholars can still at times find to disagree upon in assessing what is now in itself known in such abundance, sometimes even at the most basic level of dating a manuscript."

In particular, Whiteside called attention to a "cryptic remark" made by Newton in 1664 to obtain the force acting on a body moving on an elliptic orbit.

> "If the body b moved in an Ellipsis, then its force in each point (if its motion in that point be given) may be found by a tangent circle of equal crookedness with that point of the Ellipsis."

As if to respond to the "no and yes" in his previous statement, Whiteside then took the following position concerning Newton's "cryptic remark":

> "Not only, however, was this remark original, but Newton's path to dynamical discovery might have been very different had he pursued it. In a Waste Book entry dated December 1664 he had already roughed out a method for constructing the centre of curvature, and so the 'quantity of crookedness' inverse to it, in an ellipse. (MP v.1 252–255) Six years later to jump ahead, he would in Problem 5 of his 1671 fluxion treatise derive the elegant result that the radius ρ of curvature at any point on a conic is proportional to the cube of the normal at the point down to the axis ... But I talk of a deduction that Newton never made till the 1690s."

Contrary to Whiteside's claim that Newton never made such a deduction until 1690, however, there is now considerable evidence based on Nauenberg's analysis of Newton's 1679–1680 correspondence with Robert Hooke, and on early manuscripts of the *Principia*, that Newton did pursue this "path to dynamical discovery" much earlier than scholars had previously thought.[7,9] Central to this path is the mathematical concept of curvature that Newton developed from 1664 to 1671 (and that Christiaan Huygens had developed somewhat earlier). Specifically, Newton used elements of the circle of curvature to represent elements of the curve generated by a given force.[10]

This curvature analysis of motion also clarifies the seminal role which Hooke played in the development of Newton's ideas on dynamics. In his early curvature approach to dynamics, Newton resolved a continuous central force acting on a body into tangential and *normal* components; the former is responsible for the changing magnitude of the velocity along the orbit, and the latter gives rise to the curvature or bending of the orbit. Hooke suggested, however, that the motion be resolved into tangential and *radial* components, and that the force be *impulsive* rather than continuous. Newton implemented Hooke's suggestion mathematically with the central force represented by a periodic sequence of impulses, and the motion resolved into tangential and radial components; the former is given by the tangential velocity, and the latter is due to the change in velocity "impressed" by the impulse. As a consequence, Newton shortly afterwards found the origin of Kepler's law of areas, which previously had been hidden in his curvature approach. Thus, he was able to implement the purely geometrical approach to orbital dynamics that is developed in the *Principia*, in which time is measured by the change of area swept out by the radial vector.

In his lecture "From Kepler to Newton: Telling the Tale," Curtis Wilson recounted that Newton regarded Kepler's laws as empirical rules, which Kepler had only "guessed" but which he, Newton, had demonstrated as a consequence of the fundamental laws of nature. Wilson described how Kepler derived his rules from Tycho Brahe's planetary observations, and how these rules were thought to be only empirical, even by Voltaire. Newton realized that these rules are strictly valid only for the ideal orbital motion about a fixed center of force, and that they are only approximate for the case of planetary motion around a free central body with mutual gravitational interactions among a number of planets. In 1684, he wrote the following in the *De Motu*, the first draft of a manuscript which later became the *Principia*:

> "It may happen that the centripetal force does not always tend towards that immobile center (the sun), and thence that the planets neither revolve exactly in ellipses nor revolve twice in the same orbit. Each time a planet revolves it traces a fresh orbit as happens also with the motion of the Moon, and each orbit is dependent upon the combined motions of all the planets, not to mention their action upon each other."

On the continuing controversy concerning the nature of Newton's Moon test for the inverse square law of gravity in the 1660s, Wilson reported that there is clear evidence in a manuscript, which was first published by Rupert Hall and dated by him before 1669, that Newton did make the test. However, some of Newton's central arguments for the universal law of gravitation (i.e. that the gravitational force acts between all matter and is proportional to the masses of the interacting bodies) became clear to him only sometime during 1685, while he was in the midst of writing the *Principia*.

John Fauvel elucidated the development and impact of Newton's calculus in his lecture on "Newton's Mathematical Language." Quoting John Maynard

Keynes, who thought that "The proofs (in the *Principia*), for what they are worth, were, as I have said, dressed up afterwards — they were not the instrument of discovery," Fauvel raised the thorny question of whether there is a gap between Newton's private mathematical language and his exposition in the *Principia*. Much light has been shed on this issue by the availability of Newton's mathematical papers. These papers reveal, as Whiteside puts it, "... that he (Newton) made mistakes, that he learned from them and that with unwearying application he steadily enlarged his grasp as he constructed the mature fluxional calculus." Newton also gave a careful analysis of the transition between natural everyday language and mathematical language, and emphasized the importance of introducing mathematical definitions carefully. Inadequate attention to this point has led to unnecessary confusion among Newtonian scholars.[11] Fauvel pointed out the little known fact that Newton first discovered infinite power series expansion of certain algebraic expressions by transforming age old arithmetic rules into an algebraic language:[12]

> "I am amazed that it has occurred to no one (if you except N. Mercator with his quadrature of the hyperbola) to fit the doctrine recently established for decimal numbers in similar fashion to variables, especially since the way is then open to more striking consequences. For since this doctrine in species has the same relationship to Algebra that the doctrine in decimal numbers has to common Arithmetic, its operations of Addition, Subtraction, Multiplication, Division and Root-Extraction may easily be learned from the latter's provided the reader be skilled in each, both Arithmetic and Algebra."

One of the most difficult questions that has challenged historians of science has been Newton's application of his perturbation methods celestial dynamics to the motion of the moon. In an article entitled "From High Hope to Disenchantment," Whiteside[13] concluded that Newton's deduction of the lunar inequalities "was a retrogressive step back to an earlier kinematic tradition which he had once hoped to transcend." However, as Michael Nauenberg showed in his contribution to the symposium, "Newton's Portsmouth Method and its Application to Lunar Theory," it is misleading to see Newton's lunar theory as a failure. In fact, Newton's methods gave remarkably successful approximate solutions to the notoriously difficult three-body problem. By these methods, he calculated the inequalities of the lunar motion (some known from antiquity) that are due to the gravitational perturbation of the Sun. Even the most striking shortcoming, the treatment of the rotation of the line of apsis as it appears in the *Principia*, is discussed by Newton in a remarkably profound manner in one of his previously unpublished manuscripts, which was edited by Whiteside.[14] Apparently, Newton was dissatisfied with his results; they never appeared in any of the three editions of the *Principia.*

In his concluding comments Rupert Hall aptly summed up the challenge and response of the symposium:

> "It is most illuminating that recent studies have begun to penetrate beneath the polished, admantine surface of Newton's great printed works to the foundations of his public formulations, to reveal the possible alternative arguments or positions that Newton considered and rejected. We now know that *Principia* and *Opticks* did not spring like Minerva from the head of Jove: they are a palimpsest of investigation and tentative endeavors."

Notes and References

1. Following Newton's death in 1727, the collection was secured by John Conduitt, passed into the possession of the Portsmouth family in 1740, and for over two centuries was virtually inaccessible to scholars save for a few determined individuals such as David Brewster and W.W. Rouse Ball. In 1872, the fifth Earl of Portsmouth, Isaac Newton Wallop, transferred to Cambridge University the portion of the collection judged to be concerned with scientific and mathematical topics. In 1936, essentially all the remaining papers were sold by Sotheby at public auction. The bulk of the theological manuscripts now reside in the Jewish National and University Library in Jerusalem, a few other lots of papers reside in private hands, but the vast majority are available to scholars in England who have the time and resources to examine them under the careful eye of university librarians. A new dimension to Newtonian scholarship was added, however, with the publication of *Isaac Newton's Philosophiae Naturalis Principia* (3rd ed. (1726) with variant readings (1972, eds. I.B. Cohen and A. Koyre, assisted by A. Whitman), *The Correspondence of Isaac Newton* (1959–1977, seven volumes, eds. Hall, Scott, Tilling, Turnbull), *The Mathematical Papers of Isaac Newton* (1967–1981, eight volumes, ed. D.T. Whiteside) and *The Optical Papers of Isaac Newton* (1984, Vol. 1, ed. A. Shapiro and Vol. 2 now being prepared by Shapiro).
2. Thomas Young, "The Bakerian Lecture: On the theory of light and colours," *Philosophical Transactions* **92** (1802): 12–48.

3. "The optical papers of Isaac Newton," in *The Optical Lectures* 1670–1672 (ed.) Alan Shapiro (Cambridge University Press, 1984), Vol. 1. Shapiro is currently editing Vol. 2.
4. M. Nauenberg, *Comparison of Newton's diffraction experiments with Fresnel's theory*, in this volume.
5. R.G. Lunnon, "Fluid resistance to moving spheres," *Proceedings of the Royal Society of London*, Series A, Vol. 110, 302–326 (1926).
6. J. Bruce Brackenridge, "The critical role of curvature in Newton's dynamics." in *An Investigation of Difficult Things: Essays on Newton and the History of the Exact Sciences* (eds.) P.M. Harman and Alan E. Shapiro (Cambridge University Press, 1992).
7. M. Nauenberg, "Newton's early computational method for dynamics," *Archive for History of Exact Sciences* **46** (1994): 221–252.
8. D.T. Whiteside, "The prehistory of the *Principia*: from 1664 to 1686," *Notes and Records of the Royal Society of London* **45** (1991): 11–61.
9. J. Bruce Brackenridge *The Key to Newton's Dynamics: The Kepler Problem and the Principia* (University of California Press, 1995).
10. Following Brackenridge's paper, Whiteside rose to express his continued opposition to the curvature analysis of Newton's 1679 diagram. Unfortunately, he has decided not to present these remarks in this volume, while stating his opinion that alternate interpretations are still possible.
11. An example is Newton's apparent change over time of the *meaning* of the term "centrifugal" force, although he was always consistent in his *mathematical application* of it to dynamical problems.
12. *The Mathematical Papers of Isaac Newton*, 1670–1673 (ed.) D.T. Whiteside (Cambridge University Press, 1969) Vol. 3, pp. 33–35.
13. D.T. Whiteside, "Newton's lunar theory: from high hope to disenchantment," *Vistas in Astronomy* **19** (1976): 317–328.
14. *The Mathematical Papers of Isaac Newton*, 1684–1691 (ed.) D.T. Whiteside (Cambridge University Press), Vol. 6, pp. 508–535.

I. Bernard Cohen

Newton's Scholarship in Historical Perspective*

I. BERNARD COHEN
Harvard University
Cambridge MA 02138, U.S.A.

In the last half-century, our views concerning Isaac Newton have undergone radical changes. Compared to the past, we have a deeper understanding of Newton's science and mathematics and we have become aware of his full creative stature and the many dimensions of his complex personality. A convenient place to begin our examination of these changes is the two-volume biography by David Brewster, published in 1855. At that time, the fashion was to write of historic personages in an adulatory mode. Brewster hailed Newton as the "High Priest of Science."[1] One of the positive features of these volumes is that they did make available some selections from the manuscripts belonging to the family of the Earl of Portsmouth, Newton's collateral descendants.[2]

Brewster's biography is notorious for the treatment of Newton's alchemy and his religious beliefs. Brewster simply could not "understand how a mind of such power, and so nobly occupied with the abstractions of geometry, and the study of the material world, could stoop to be even the copyist of the most contemptible alchemical poetry." How could a Newton take seriously and annotate the *De re metallica* of Agricola, which Brewster declared to be "the obvious production of a fool and a knave."

Although aware of Newton's unpublished writings on theological questions, Brewster either did not see or purposely ignored Newton's statements of disbelief in the doctrine of the Trinity. He praised the "wise discretion" of "Doctor Horsley," the eighteenth century editor of Newton's *Opera*, who chose not to make public Newton's statements about his religious beliefs.[3]

In retrospect, what is most disappointing about Brewster's biography, however, is not the attitude toward alchemy and religion, but rather the failure to illuminate our understanding of Newton's actual science and mathematics. Thus, in today's world, this work is not very useful to scholars and pales by comparison with the meticulously edited and copiously annotated edition of the Newton-Cotes correspondence, produced by J. Edleston. This work was

*This paper is dedicated to D.T. Whiteside and to Rupert and Marie (Boas) Hall, for their great contributions to our knowledge of the scientific thought of Isaac Newton.

first published in 1850 and it is still today a useful and important tool for Newton scholars.

Until the latter part of the nineteenth century, a huge collection of manuscripts by or relating to Newton was still in the possession of Newton's collateral descendants. These consisted of Newton's own records of correspondence (both letters to and from Newton), drafts of major and minor works, various kinds of essays on many different subjects, together with the documents assembled by John Conduitt (husband of Newton's niece) for a planned biography of Newton. These papers were inherited by the Conduitts' only child, who married Viscount Lymington, whose son was the second Earl of Portsmouth. This collection, generally known as the "Portsmouth Papers," was kept in Hurstbourne Castle until the Earl and his family decided that Newton's scientific manuscripts would be better preserved in some public repository. Accordingly, in the 1870s, what was called the "scientific portion" of the manuscripts was deposited in the University Library in Cambridge.

This enormous collection, rich in all kinds of materials relating to Newton's life — his thought, his work in mathematics, physics and astronomy etc. — was described in a published catalogue (1888), prepared by a University syndicate, whose members included the mathematical physicist G.G. Stokes and the astronomer John Couch Adams.

This catalogue is notably non-revealing. The descriptions are brief, laconic to the extent of being useless, hardly revealing either the quality or the extent of the items in question. One of the positive features of this catalogue was an introductory essay, of which a portion was obviously written by Adams, who revealed some of the astronomical and mathematical treasures he had found, including materials on the lunar theory and the solid of least resistance.

Over the next decades, this extraordinary hoard of Newtoniana attracted little scholarly attention and was hardly used. One of the few who even deigned to look at any of these manuscripts was W.W. Rouse Ball, known today chiefly for his popular book on "mathematical recreations."[4] His books did not make the world cognizant of the great treasures awaiting study in the University Library.

Between the time of the gift of the Portsmouth Papers and the 1930s, there were a few others who did make some use of this vast collection. The work of these scholars, however, did not declare to the world at large the importance of studying Newton in the original sources. In trying to understand why this was so, we must remember that in those decades there was as yet no real discipline of the history of science and of mathematics. Much of the writing in these areas was then produced by scientists or mathematicians in their retirement or scientists indulging in a hobby. The number of individuals who produced lasting historical contributions in the history of science and mathematics was small, including such heroic figures as J.L. Heiberg, G. Eneström, Thomas Little Heath and Paul Tannery.

An index of the lack of response to the existence of great source materials in the history of science and mathematics is the relatively small use made of the great edition of the "Oeuvres" of Christiaan Huygens. In terms of completeness of edited texts — correspondence, manuscript drafts and notes, completed published works — there is no other edition of the papers of a scientist or mathematician comparable with this one. And yet in the 1920s, 1930s and 1940s, there was not produced an adequate full biography of Huygens and scarcely any noteworthy studies on Huygens's science and mathematics.

In 1934, Louis Trenchard More, Dean of the Graduate School of the University of Cincinnati, published a 675-page biography of Newton. Even though More did make some use of manuscript sources, quoting or citing many hitherto unnoticed documents from the Portsmouth papers (from Cambridge as well as from those still remaining in the possession of the family), his book did not spark a new interest in Newton, either in the personality of this extraordinary man or in his scientific work.

There are a number of reasons why this was so. First of all, the times were not as yet ripe in terms of scholarship in the history of science and of mathematics. More importantly, More's book is dull, incident following incident and topic following topic without much illumination. Newton is generally treated with such veneration that the real person scarcely emerges.

Additionally, More was an extreme arch-conservative in science. Thus, after lauding Newton's *Principia* as a work of equal greatness with Aristotle's *Organon*, he deplored the fact that "these two works, probably the two most stupendous creations of the scientific brain," were both "now under attack." The enemies were, respectively, "the relativists in physics," led by "Professor Einstein," and what he called "modern symbolists in logic" — a curious name for those developing mathematical logic or, as it was then often called, symbolic logic. More insisted that "Aristotle and Newton will be honoured and *used* when the modernists are long forgotten."[5]

More does deserve credit, however, for having stated frankly and clearly the nature of Newton's beliefs on such doctrinal questions as the Trinity and for having published extracts of Newton's expressing his anti-Trinitarian beliefs. Yet, More did not provide any real illumination and understanding of Newton's science and mathematics. The almost 700 pages of this work do not contain a single diagram nor so much as a single equation or proportion.

As all Newton scholars are aware, the big event that almost changed the availability of Newtonian sources overnight was the sale at Sotheby's public auction in 1932 of the vast horde of Newton papers still belonging to the family of the Earl of Portsmouth. This sale was provoked by the need to satisfy the payment of death duties.

The sale dispersed Newton's papers to the far corners of the earth. There was a notable scholarly catalogue produced for the sale. It contained many generous extracts from documents hitherto unknown or inaccessible to scholars. The descriptions of the various manuscripts and other items put up for auction, together with copious extracts, revealed more about Isaac Newton as a person and about various aspects of the development of his scientific thought than More's biography of a few years earlier.

Prior to the late 1940s, however, the only reference in print that I know of, in which any use was made of these newly available manuscripts is a curious paper by John Maynard Keynes. Keynes, as is well-known, assembled a considerable mass of the Newton papers disseminated in the Sotheby sale; these are now in King's College Library. In particular, Keynes owned a number of manuscripts dealing with alchemy and with theological subjects.

On the basis of an examination of these manuscripts, Keynes produced a paper, first read to a group at Trinity College and later to the Royal Society Club. Keynes was no longer living when in 1946, the Royal Society somewhat belatedly celebrated the 300th anniversary of the birth of Isaac Newton. His brother Geoffrey — the surgeon, bibliophile and book collector — read this paper at the celebratory meetings, and it was published in the Royal Society's volume.

Entitled "Newton the Man," this essay has become famous for its radical portrayal of Newton. Keynes insisted that his reading of Newton's manuscripts revealed a Newton who was not "the first and greatest of the modern age of scientists, a rationalist, one who taught us to think on the lines of cold and untinctured reason." Rather than being "the first of the age of reason," Keynes presented Newton as "the last of the magicians," the "last wonder-child to whom the Magi could do sincere and appropriate homage." In short, Keynes denigrated Newton's mathematics and physics and his founding of celestial dynamics, denying that he had been "the first and greatest of the modern age of scientists."

The new Newton was to be considered "a magician" who "looked on the whole universe and all that is in it *as a riddle*, as a secret which could be read by applying pure thought to ... certain mystic clues which God had laid about the world to allow a sort of philosophers' treasure hunt to the esoteric brotherhood." In today's scholarly world of irrationalism and lack of respect for any realism in scientific truth, Keynes' statements about Newton as a non-rational magician find an ever-growing sympathetic audience.[6]

As far as I know, the next work to be based on the Newton manuscripts was a pair of scholarly articles published by Rupert Hall in 1948 and 1957. The first of these dealt with one of Newton's early notebooks, but the second one, entitled, "Newton on Central Forces," used manuscript sources to explore the genesis and development of Newton's concepts in dynamics. Despite the fact that Hall had shown how new insights into the development of Newton's scientific thought could be found by studying the manuscripts in the University Library, his example was not immediately followed by others. Indeed, I think it fair to say that it was not until Rupert and Marie Hall published their volume, entitled *Unpublished Scientific Papers of Isaac Newton*, in 1962 that the scholarly world at large, the world of historians of science and of scientists and mathematicians interested in historical questions, became aware of some of the extraordinary new insights that were to be found by examining the unpublished Newton materials.

Of course, by this time, in 1962, several others of us had already begun our projects of using Newton manuscript materials. The Royal Society had undertaken the edition of the *Correspondence* of Isaac Newton, of which the first volume appeared in 1959, three years before the volume of *Unpublished Scientific Papers.* I myself had begun, in close collaboration with Alexandre Koyré, to study Newton's manuscripts and annotated books in order to learn about the genesis and development of the concepts and methods of his great *Principia.* Some others, notably John Herivel (in 1959), had begun to publish the results of explorations in the Newton manuscripts.[7] Even more important, of course, was the enterprise of D.T. Whiteside, eventually resulting in the eight magnificent volumes of the *Mathematical Papers of Isaac Newton* (of which the first volume appeared in 1967).

It is difficult to think of any work of scholarship of comparable magnitude produced in our time, any other which matches the stature of this tremendously important work. For not only do we find here, in full display, the documents that mark the development of Newton's thinking in mathematics, but there are also extensive annotations and commentaries which provide more information concerning the development of the exact sciences in the seventeenth century than is available in most treatises. In short, this work — by any conceivable standard — is a landmark. In retrospect, what may be most remarkable about the *Mathematical Papers* is that this was the production of a single human being working without any assistants, save at the stage of proofreading, and that the whole job was completed in twenty-two years!

It should be noted that Rupert Hall's contribution to our understanding of Newton did not cease with the volume of *Unpublished Scientific Papers*, which he produced with his helpmeet Marie. Rupert and Marie produced together a number of scholarly papers, of which those on Newton's chemistry and alchemy are especially significant and not always fully appreciated. Furthermore, it was under Rupert Hall's direction and editorship, assisted by Laura Tilling, that the eight-volume edition of Newton's correspondence was finally brought to fruition. And it is not amiss to mention here, Rupert's later volume on the Newton–Leibniz controversy and his relatively short biography of Newton (short, when compared to Sam Westfall's monumental work), and his more recent study of Newton's optics.

Since the beginnings made by Tom Whiteside and Rupert Hall, many important sources have been produced for the use of Newton scholars, too many to be listed or even mentioned here. But I must take note of the Westfall biography, *Never at Rest* (1980), a fabulously rich resource book for almost any aspect of Newton's life and thought. And I am sure that everyone here will wish godspeed to Alan Shapiro's projected further volumes of his important edition of the *Optical Papers*, a scholarly endeavor of the highest order which commands the respect of Newtonian scholars. One of the purposes of our gathering here today is to pay tribute to Tom Whiteside and to Rupert and Marie Hall and to display some of the results of Newton scholarship that have made use of manuscript sources of various kinds — sources that were virtually unknown to scholars prior to the scholarly work of Tom Whiteside and the Halls and all the others who have followed in their footsteps.

As I have just said, I know of no work in the history of mathematics and the sciences that is on par with Tom Whiteside's edition of the *Mathematical Papers*. Of course, whenever one thinks of monuments in the history of science, that which leaps immediately to mind is the magnificent edition of the works of Christiaan Huygens to which I have referred to earlier. But Tom's edition of the *Mathematical Papers* goes beyond the scope of the Huygens edition in one very significant feature: almost every item presented in *Mathematical Papers* is accompanied by Tom's extensive annotation and commentary — not only on strictly textual problems and the general historical aspects of the document (that is, when and where and why it was written, how it was received, and so on). Tom's running commentary, often with mountains of erudition packed tightly in a telegraphic style into a small compass, is in a class by itself — the equivalent in information and explanation of a whole library of scholarly books and articles. These historical commentaries and technical explanations or glosses are of enormous help to any reader who really wants to understand what Newton was doing and why he did so.

I know of no other mathematical or scientific texts that have been so completely annotated in a modern scholarly edition as these "papers" of Newton's. Recently, I had occasion to re-read some parts of Descartes's *Géométrie*, which — as many of you will know — is not the easiest of books to read. As I was studying this seminal text, I could not help but wish that some Tom Whiteside had produced an annotated edition on the scale of the *Mathematical Papers* to guide ordinary readers through the text and to help them understand the implications of each part.

How was Tom Whiteside able to produce so magnificent a tool for scholars? As I see it, he came to the project with several qualifications which in combination were needed for this assignment. First of all, he began with an unrivalled background knowledge of seventeenth century mathematics. He had steeped himself in this early mathematics while serving his time as a research student at the University of Cambridge and the evidence is available for all to read in the published version of his doctoral dissertation, "Patterns of Mathematical Thought in the 17th Century." Let me say, as an aside, that when I first got to know Tom he was just finishing his dissertation. At that time, Clifford Truesdell, the founder and editor of *Archive for History of Exact Sciences*, had appointed me as a member of the advisory board and so I had the honor (along with Carl Boyer) of recommending this work for publication. Truesdell later told me that he considered this to be the most important contribution he had published in the journal.

The second quality that Tom brought to the Newton project was his training in languages and the study of texts, a feature of his primary field of study (Romance languages) while an undergraduate at Bristol. A knowledge of language and the ability to deal with expressions in foreign languages is a gift, but one that is honed by training. Tom not only came to the Newton project skilled in languages and able to read Latin, French, and German, he also had been introduced to the integrity of texts.

Although Tom Whiteside learned much about the nature of texts and text-editing from the works he studied at Bristol, he found the mode of studying texts boring to him. Yet he made his way dutifully through such authors as Vergil, Caesar, Cicero, Horace, Martial and, his favorite "Golden Age" author, Tacitus, along with Villehardouin, Rabelais, and Montaigne. He found the prescribed study of Latin and French authors so uninteresting that his active mind sought for another source of stimulation, which he found in reading books on the history of mathematics, including important works by T.L. Heath and O. Neugebauer and ver Ecke's edition of Pappus. As he was not adhering to the orthodox mode of study, he even came close to losing his honors status, which would mean forfeiting his state scholarship. Thanks to the creation of a new university general honours degree in Arts, Tom was able to combine work in languages and literature with logic. Cramming for the final exams by reading the French books he had not yet studied, he was able to do well enough for a good degree and at the same time exhibited his command of Latin prose by making a version in the style of Tacitus of a set piece in English. Later in life, Tom recalled that he had actually purchased and enjoyed using a French edition of Lucretius which had the Latin original and the French translation on facing pages, with the editor's annotations at the bottom of the page. This is the form he later adopted in his edition of Newton's *Mathematical Papers*, but he does not remember making a conscious decision to adopt the form used by this French scholar.

A third quality, or qualification, was an extraordinary gift to think mathematically. Repeatedly in the *Mathematical Papers*, as in other publications, Tom displays a command of mathematics itself that is all the more astonishing in one who for whom mathematics was not the major professional subject of study in college or in graduate school. Tom recalls that at Bristol he never took a formal course in mathematics, but he read passionately many works on the history of mathematics and worked up various topics in mathematics on his own. One subject that especially delighted him was algebraic geometry.

A striking example of Tom's mastery of mathematics may be found in his reconstruction of the steps by which Newton produced a new proof of Proposition 10 of Book II for the second edition of the *Principia*. Newton's attention had been called to a fault in this proof, but only after this portion of the *Principia* had already been printed. Accordingly, Newton not only had to produce in short order a correct proof to replace the faulty one; he also had then to pare the new proof until it could fit in the available space. Only someone skilled in mathematics could fully understand the basis of the criticism of the original Proposition 10. Furthermore, because Newton was, on this occasion, as on so many others, a veritable pack-rat, saving every scrap of his development of the new proof, Tom Whiteside had to face the formidable mathematical task of figuring out the chronological order of these scraps and versions and then understanding Newton's mathematical procedure. Here his skill as mathematician served him well.[8]

My fourth quality that enabled Tom to produce the edition of *Mathematical Papers* is a gift for the study of handwriting. This has long served him in making transcriptions of Newton's manuscripts, but has also enabled him to date various drafts and versions with an astonishing degree of accuracy — tested on many occasions by later corroborative evidence. In my own work, again and again, I found Tom's ability to date Newton's manuscripts by the handwriting to be of enormous help. And he equally was of great help in identifying the handwriting of others for example, in the annotations found in various copies of the *Principia.*

There are yet other qualities that were needed in order to produce the edition of *Mathematical Papers.* One, of course, was an incredible gift of memory, the ability to remember and to keep in mind the individual parts of the vast number (literally thousands) of different manuscripts, so as to be able to recognize that a given document is related to another in a very different part of the collection. Of course, the most important quality of all is the gift of historical-mathematical insight, that undefinable aspect of the mind that shines through every part of this great work.

In thinking about ways to illustrate how Tom Whiteside's work has made a difference to our knowledge of Newton, or indeed of our image of Newton or of our understanding of Newton's thought, I have chosen two very different examples. In many ways, for me the most dramatic finding of Tom Whiteside is the importance of Descartes in the early formulation of Newton's mathematics. Before the edition of *Mathematical Papers*, our knowledge of this subject was based on what seemed to be irrefutable evidence of Newton's disdain for Descartes. There was the statement recorded by Pemberton that Newton regretted having begun his study of mathematics by reading the moderns rather than the ancients. This was interpreted as indicating his regret at wasting time with Descartes's *Géométrie* rather than studying Euclid and Apollonius.

The second was the statement by Brewster that Newton's copy of Descartes's *Géométrie* was marked throughout "Error. Error. Non Geom." For some time it was thought that this might have been an exaggeration by Brewster, since there seemed to be no such copy in existence. But finally, in part due to Tom, this puzzle was resolved. Newton did make such marks in a copy of Descartes.

When such evidence concerning Newton's attitude toward Descartes was coupled with the conclusion of Book II of the *Principia*, the conclusion seemed certain: Descartes was not one of those who had exerted a formative influence on Newton.

But Tom changed that view entirely in so far as mathematics is concerned. He showed how Newton's early views of the calculus were forged while making a close study of Descartes's *Geometry*, not the edition in French in which he gleefully noted the errors, but the edition in Latin with the commentaries of Frans van Schooten and others. This introduced Newton not only to the mathematical concepts and methods of Descartes himself, but also the important innovations of the Dutch school who were van Schooten's pupils, notably Hudde and Huygens.

The revelation of this seminal role of Descartes, enriched by van Schooten, was paralleled by the recognition by Alexandre Koyré, at more or less the same time, of the ways in which some of Newton's concepts concerning motion were also conditioned by his reading of Descartes. Koyré showed us how Newton took from Descartes the concept of "state" of motion or of rest, and developed Descartes's ideas in his own formulation of Definition 3 and Law 1. I myself was able to confirm this particular indebtedness. Today, we are aware that the "*Axiomata sive leges naturae*" of Newton's *Principia* were a kind of transformation of what Descartes called "*Regulae quaedam sive leges naturae*" in his *Principia*.

This shift in the evaluation of the role of Descartes in Newton's thought was a dramatic one. I clearly remember a troubled colleague asking me how it was that scholars were now discussing the importance of Descartes in the development of Newton's thought. Like others of his generation, he had always thought that the subject of Newton and Descartes was epitomized by the second book of the *Principia*. Here, it will be recalled, Newton concludes by showing that vortices cannot exist, that the celestial system of Descartes contradicts Kepler's laws of motion. Alexis Clairaut even went as far as to suggest that the sole purpose of Book II of the *Principia* was to destroy Descartes's system of vortex-based cosmology.

A second revelation made by Tom Whiteside was in his gloss on Proposition 41 of Book I of the *Principia.* Until recently, most scholars had limited their study of the *Principia* to the definitions and laws and then the first three sections of Book I. They skipped all the rest of Book I and also the whole of Book II, finally studying the first part of Book III and the concluding General Scholium. Indeed, it will be recalled that this is in fact Newton's own suggestion to readers in the beginning of Book III.

Today however, it is becoming generally recognized that we do not really begin to see Newton as the master of mathematical physical science until we get further along in Book I. Here we needed someone like Tom to guide us, to make clear, at the start, the exact nature of Newton's dependence on what he called "the quadrature of certain curves," or the ability to perform the integration of certain functions. Even more important, Tom's guidance made it evident that in Proposition 41, as elsewhere, Newton was actually making use of the calculus and that his very language, when read carefully, permits of no other reading. For example, when Newton writes about "the line element IK" which is "described in a given minimally small time," he clearly has in mind what we would call a distance *ds* described in a time *dt*. In short, the cluster of propositions around Proposition 41 are written in the language of the differential and integral calculus and thus their analytic character is only partially masked by their synthetic form of expression.

Our meeting today at the Royal Society is a celebration of two great contributions to Newtonian scholarship — by Tom Whiteside and Rupert Hall — notably their heroic efforts in producing and interpreting Newtonian and Newton-related texts on which all Newton scholars depend. I have mentioned earlier the seminal work of Rupert Hall, much of it done in concert with Marie, in revealing the extent of Newton's thought beyond the works he allowed to appear in print. Anyone engaged in Newtonian research, and indeed, in research on any aspect of the science of the seventeenth century, depends heavily on the extraordinary edition of the correspondence of Henry Oldenberg, first Secretary of the Royal Society, an edition produced by Rupert and Marie Boas Hall.

I have already mentioned Rupert's bringing to a happy conclusion the Royal Society's edition of the *Correspondence* of Isaac Newton and the stream of important scholarly articles produced by Rupert and Marie.

Earlier in this presentation, I have said that for most scholars, a real turning point was the volume of *Unpublished Scientific Papers*, which was edited by Rupert and Marie Hall. Let me demonstrate the seminal importance of this work by indicating three ways in which its contents have changed our general thinking about Newton's ideas. Of course, the primary quality of this work is that Rupert and Marie actually found, identified, and interpreted, a large number of documents, the very existence of which was then unknown to the scholarly world at large. The interpretive essays and commentaries by the Halls explained the significance of the new documents they had found and were presenting.

In retrospect, what may have been the most notable feature of *Unpublished Scientific Papers* was the editing and translating of Newton's essay, beginning "*De gravitatione et aequipondio fluidorum* ...." The Halls recognized the significance of this document as Newton's response to a first contact with Descartes's *Principia.* Here we find Newton formulating major concepts concerning space, time, and motion, and also force and inertia, in a Cartesian framework, just as was the case for his mathematics. The publication of this essay, together with the Halls' introductory commentary, documented Newton's first full encounter with the philosophy of Descartes and wholly changed our idea concerning the influence of Descartes on Newton. The late Betty Jo Dobbs recently published her study of this manuscript and had concluded that it dates from a time just before the *Principia* rather than earlier. I myself have found corroborating evidence for this re-dating.

A second important contribution of *Unpublished Scientific Papers* was the publication of a set of papers relating to Newton's early thoughts about motion. These include what is, I believe, the first English translation made of the tract *De Motu.* Although this tract was published at least twice before, in the original Latin, the Halls not only gave the first English translation but also listed the various extant versions and gave extracts indicating the differences among them. Thus, for the first time, this earnest of the great *Principia* to come was made available in a form for general use by scholars.

Finally, a third revelation — and in some sense the most important of all — was the existence of preliminary versions of an introduction and conclusion planned for the first edition of the *Principia*, including some thoughts that finally appeared in the second edition of the *Principia* in the concluding General Scholium and in later Queries of the *Opticks.* The Halls also found documentary fragments relating to these rejected texts.

Until the Halls published their volumes, one of the great scholarly puzzles for students of Newton had always been that the first *Principia* ends abruptly in a discussion of comets. How could Newton have written so magisterial a work on Natural Philosophy without a proper conclusion? The Halls found the answer. He had planned a General Conclusion in which he would indicate, *inter alia*, how his findings might be extended into other domains of natural philosophy, suggesting how his findings in rational mechanics and celestial dynamics might be carried into studies of the physics of matter, notably the constitution of matter and the action of short-range forces between constituent elements of matter. Suggestions about how to explore these topics and other related subjects were also explored by Newton in the manuscript Preface which the Halls discovered, edited, and published, and which, like the abortive Conclusion, was also rejected by Newton and never completed. In retrospect, we can see that Newton was acting wisely in his decision not to burden his *Principia* with debatable and intimate speculations. There were enough topics which were bound to arouse hostility, such as the introduction of a gravitating force "acting at a distance," a concept abhorrent to all scientists who were followers of the reigning "mechanical philosophy."

There was another long-standing puzzle about the *Principia* which the Halls were able to solve. In the eventual General Scholium, which appeared for the first time in the second edition of the *Principia* (1713), a final paragraph discusses what Newton calls a "spiritus" or spirit. What did he mean? No one was quite sure. Professor Koyré even hazarded a guess that Newton may have had in mind the "spirit of God." But the Halls solved this puzzle by finding some preliminary drafts, which they published for the first time. They discovered that the "spirit" which Newton had in mind was an aspect of the new science of electricity then being developed by Francis Hauksbee, whose relation with Newton was later the subject of several important monographic studies by the late Henry Guerlac. The Halls clearly showed that the kind of "spirit" Newton had in mind in the General Scholium was that of an "electrical" spirit, the active agent producing the various effects observed by Hauksbee and demonstrated to Newton at meetings of the Royal Society.

Before leaving the subject of the contributions of the Halls and of Whiteside, let me add a final word about paths of influence. I have stressed the production of texts and their interpretation by both Tom Whiteside and by Rupert and Marie Hall. But these three researchers have influenced generations of scholars by means of their published articles. Scholarly influence is also produced by personal contact, by a laying-on of hands. At Indiana and later in London, the Halls have trained many young scholars and have also influenced many senior scientists and historians who have worked with them on research in the history of science. It should also be mentioned that Rupert later went on to become an important pioneering figure in the new discipline of the history of technology and Marie produced valuable and interesting studies of the history of the Royal Society.

I first got to know Rupert Hall in the 1950s, when I came to Cambridge to work on the Newton papers. He was then still a Fellow of Christ's and had already begun to publish articles about Newton. As he and I contemplated the vast store of information lying unread and unused in the various Cambridge repositories (remember this was before the edition of the *Correspondence* or of the *Mathematical Papers*) we discussed a joint endeavor which we envisaged as a two-volume work on *The Life and Times of Isaac Newton.* Obviously we never got around to producing such a large-scale work, but each of us did publish a shorter biographical account. Mine appeared in the *Dictionary of Scientific Biography* and his as a separate book, entitled *Isaac Newton — Adventurer in Thought.* The volume *Never at Rest*, the magisterial biography written by our friend, the late Sam Westfall, approaches our planned work in its grand scale.

I should like to conclude by mentioning a feature of Tom Whiteside's contribution to Newton scholarship. Tom has been a teacher in a number of different ways. Those who have been able to attend his Cambridge lectures on the history of mathematics may be envied their good fortune. But Tom has also been a teacher in other ways as well, primarily by being a most generous advisor to many Newton scholars who have sought his help. Some have gained knowledge of sources they were unaware of and others have profited from answers to specific questions. Many of us has have learnt much from the extensive commentaries (some running to thirty pages or more of handwritten text) that he has written about drafts of our articles or books. His generosity toward other scholars is legion and is balanced by a sharp critical sense that comes into action whenever he encounters shoddy work or pure bluff.

I cannot even count the number of occasions on which Tom Whiteside has been generous to me, in providing me with information about sources which I might not have known, or in correcting misapprehensions, or even in the giving of good scholarly advice. When Anne Whitman and I were doing our Latin edition of the *Principia* with variant readings, he (and also Mr Adolf

Prag) read the proofs of our variant readings, thus helping us to eliminate errors. When Ann Whitman and I were finally dragooned into producing a new English translation of the *Principia*, Tom's advice to us was of enormous importance and we followed his general suggestions for our procedure.[9]

Max Planck once wrote that he had had two great honors in life. One, of course, was to have made the fundamental discovery of the quantum of energy. The other, he declared, was to have been associated with Albert Einstein. I believe we can all be equally proud of the honor of having known, or having been influenced in our scholarly understanding, by Tom Whiteside and by Rupert and Marie Hall. Of them and their work, it can be said in the words of the statement concerning Sir Christopher Wren in St. Paul's Cathedral, "*Si monumentum requiris, circumspice.*" The greatest tribute to their scholarly work is to be found in the studies based on their foundations, of which a generous sample is provided in today's proceedings.

Notes and References

1. This phrase, "High Priest of Science" was lifted by Brewster from the eighteenth century accounts of Newton written by William Stukeley, an antiquary who had actually conducted a sort of oral history interview with Newton.
2. It was long thought that Brewster based his biography on a complete examination of these manuscripts, that he himself had made the selection of those he published or mentioned. The researches of D.T. Whiteside, however, revealed that Brewster had not actually had free access to all the Newton papers, but had only certain selections made for him by a younger son.
3. Almost a hundred years later, Louis Trenchard More, the next major biographer of Newton, who actually published texts of Newton's showing his unquestioned Unitarian beliefs, criticized both Horsley and Brewster, adding that there "is nothing in his [i.e. Newton's] life so serious that it should have been suppressed." I have always found this statement to be odd, since it implies that a biographer *should* suppress things in the life of the subject that are "serious".
4. Rouse Ball made some use of Newton's manuscripts in his *Essay on Newton's Principia*, publishing a transcript of a version of Newton's essay "De Motu," previously published by Rigaud from the copy in the Royal Society, and also some samples of correspondence. In another volume, called

Cambridge Studies, he included an essay by Newton on a plan for education in the University.

5. L.T. More also devoted more than a page to defending Newtonian physics against "Professor Einstein" and the "relativists".
6. This paper is a curious production. For example, there is no specific reference to a single manuscript or text, to any particular document. Furthermore, there is not a single quotation from Newton. In considering this paper, we should keep in mind that this was not a scholarly contribution written by Keynes as a serious well thought out essay intended for publication.

   Keynes wrote it for a Cambridge audience, who would welcome a presentation that was brilliant, challenging, daring, and unorthodox. It is hardly the kind of serious production that one would normally expect as part of a celebration of Isaac Newton.
7. In the Preface to *Unpublished Scientific Papers*, Rupert and Marie Hall call attention to a bare handful of scholars who had used the Newton manuscripts. The list is short, but it does include a beginning made by Alexandre Koyré and the first essays of John Herivel, whose later volume of texts is regularly cited by Newton scholars.
8. The extent of this labor can be fully appreciated only by a study of the more than a hundred pages of texts — version and emendation after version and emendation — identified, transcribed, and edited by Tom Whiteside, and explained by him in a book-length gloss on these documents (for which see *Mathematical Papers*, Vol. 8).
9. In particular, I will always be grateful to him for advising me not to worry lest our version (published in 1999 by the University of California Press) contains some (inevitable) errors that might arise because we did not fully understand every possible point in this large and complex work. Tom told us to go ahead and not to be concerned, to print our version "warts and all." Critics, he said, will call attention to errors which can then be corrected in a later edition. Another valuable piece of advice was that we should not compare our own versions with Motte's or with any other translation until we had made a complete translation of our own and revised it fully.

Alan E. Shapiro

Newton's Experimental Investigation of Diffraction for the *Opticks*: A Preliminary Study

ALAN E. SHAPIRO

School of Physics and Astronomy, University of Minnesota
116 Church Street S.E., Minneapolis MN 55455, U.S.A.

Through much of 1691, Newton had carried out an intensive experimental investigation on diffraction or, as he called it, "inflexion". Sometime around the fall of 1691, he had completed the *Opticks* and concluded it with a new part, Book IV, Part II, based on this recently completed investigation.[1] Within a few months, though, he became dissatisfied with this section and removed those pages from the manuscript of the *Opticks.* From the surviving papers for the *Opticks*, we can see exactly what went wrong. He had carried out this research with a model of diffraction that assumed that the paths of the fringes were identical to or coincided with the rectilinear paths of the rays that produced them. When he carried out an experiment that conclusively showed that this assumption was mistaken, he realized that he had to all but start over in his search for the laws underlying diffraction. At this stage of his life — the end of his active scientific career, as it turned out — he was either unwilling or unable to do this, and the *Opticks* languished incomplete until late 1702 or 1703, when he finally rewrote the brief section on diffraction and added the queries. Newton's unpublished papers on diffraction allow us to trace the path of his research and provide an unusually detailed look into the way he carried out an experimental investigation and utilized his data and calculations to deduce and reject laws.

When Newton began his investigation of diffraction in about 1691, he possessed mostly second hand and confused knowledge of diffraction, which involves the bending of very narrow beams of light into and away from shadows, thereby appearing to violate the law of rectilinear propagation, and the formation of alternating colored and dark bands or fringes alongside and within these shadows. Francisco Maria Grimaldi's discovery of diffraction was published posthumously in 1665, and it is unlikely that Newton ever read it. Newton learned of diffraction from a lecture by Robert Hooke at a meeting of the Royal Society on 18 March 1675 and from Honoré Fabri's *Dialogi physici.* The two authors presented different aspects of the phenomenon. Following

Hooke, Newton initially stressed the bending of light into the shadow and, following Fabri, the appearance of the colored fringes. Newton had written briefly on diffraction in various writings before 1691, but all he offered were vague, qualitative, and often physically impossible descriptions together with some speculations on its cause.[2]

When Newton became serious about investigating some facet of nature, he typically set out to measure it and to subject it to laws (ideally mathematical), which is precisely what he did with diffraction. The earliest drafts for the *Opticks*, reflect genuine observations and experiments, but there was no serious attempt at measurements. Newton clearly intended to make them, for he left gaps to insert numbers. As real numbers appear, so does a firmer grasp of the phenomenon.

Newton first attempted to write the new part on diffraction after he had completed what were then Books I and II on his theory of color and the three parts of Book III on the colors of thin plates.[3] He had intended his account of diffraction to directly follow the part on the colors of thin plates, but he discovered an entirely new phenomenon, the colors of thick plates, and launched an investigation of it. These two new investigations would form the concluding Book IV of the *Opticks* in its first completed state, before Newton dismembered it; the part on thick plates was Part I, and that on diffraction Part II.[4] These investigations seem to have occupied Newton from the beginning of 1691, or possibly a few months earlier, through the summer or fall of that year, when he completed Book IV. All of Newton's early descriptions of diffraction involved a single experiment, which had been described by Hooke, namely, that of a narrow beam of light passing by a knife edge. In his first draft on diffraction for the *Opticks*, he introduced a new experiment on the fringes formed by two parallel knife edges, and he proposed an innovation that allowed him to determine the distance at which the rays forming each fringe pass the knife edge.

Newton stuck the points of two knives with straight edges K in a board so that the blades formed a V or wedge that made an angle of about half a degree (Fig. 1). Placing the knives in a narrow beam of sunlight that passed through a hole H with a diameter of $1/42$ inch, he observed a series of fringes on a screen S. Three fringes parallel to each knife edge appeared (Fig. 2), and as the blades approached one another the fringes intersected. The distance between the blades was easily determined by measuring the height at which the rays passed between the knives, and for rays passing through the middle of the slit,

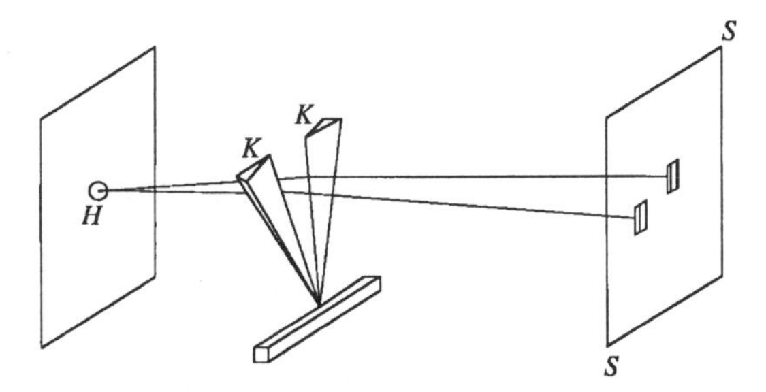

Fig. 1 A schematic diagram of Newton's experiment to observe the diffraction pattern produced by two knife blades making a small angle with one another to form a V or a wedge.

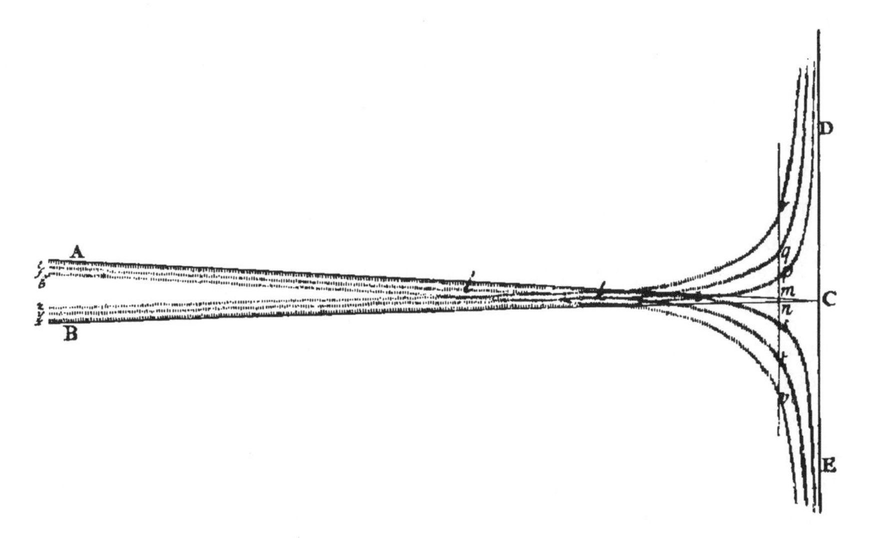

Fig. 2 The diffraction fringes formed by light passing through two knives whose edges form a V. The solid line represents the shadow of the knife edges. From the *Opticks*, Book III, Fig. 3.

the distance of the rays from the edge was simply half that distance. Newton argued that when each of the three fringes intersect, the rays pass through the center of the slit. By measuring the heights at which various colors of various fringes intersected, he was able to come up with a law:[5] "the distances of the edge of the knife from the rays which go to the same colours of the several fasciae seem to be in arithmetical proportion of the number[s] 3 2 & 1." Let s be the distance of the rays from the edge of the diffracting body, then this simple linear law can be represented as

$$s \alpha 1, 2, 3 .$$

Many of the essential features of Newton's work on diffraction for the *Opticks* can be seen here. First, and most fundamental for the suppressed version, the assumption that the paths of the rays and the fringes are identical, rectilinear paths. Second, measurements have replaced his earlier qualitative descriptions. Third, his quest to describe the phenomenon mathematically. Even though Newton left gaps in his manuscript for his measurements, I have no doubt that he had already made them and that he intended to enter refined ones later. A still more refined set of measurements would soon lead him to adopt a new law and abandon the simple linear law. The quest for mathematical laws underlying phenomena — and their continual refinement — is characteristic of all Newton's optical work. Finally, we can see that he assumes that diffraction is an instance of action at a distance. In a somewhat later draft, he formulated this assumption as a proposition:

> The surfaces of bodies act regularly upon the rays of Light at a distance on both sides so as to inflect them variously according to their several degrees of refrangibility, & distances from the surface. The truth of the above Proposition will appear by the following observations.[6]

There can be little doubt that one of Newton's aims in his investigation of diffraction was to find evidence for the existence of short-range forces between light and matter, a suggestion he had already made in the *Principia*, Book I, Proposition 96, Scholium.

In his next draft, Newton began a new series of observations of the diffraction pattern produced by a hair. Illuminating a hair with a narrow beam of light admitted through a small hole about 12 feet away, he observed three colored fringes on each side of a central shadow, which was much larger than what is predicted by the laws of geometric optics. This observation marks Newton's recognition that shadows are larger than they should be, for previously, he had followed Hooke and emphasized the bending of light into the shadow. After cursorily describing the fringes, he abruptly ended this draft (Add. 3970, ff. 375–376) and began systematic observation and measurement of the fringes.

In a new draft devoted primarily to diffraction by a hair (Add. 3970, ff. 377–378, 328–329), he set forth the results of this new investigation. The outermost fringes were found "so faint as not to be easily visible." The colors of all the fringes were difficult to distinguish unless they fell obliquely onto a screen so that they were much broader than when they fell perpendicularly. Table 1 from Newton's manuscript gives his measurements of the fringes in fractions of an inch when the screen was placed at half a foot and nine feet from the hair.[7]

Newton used the term "fascia" rather than "fringe" until about 1703, when he adopted the new term in preparing the *Opticks* for publication. We should also note that he was unable to measure directly $\frac{1}{360}$ or $\frac{1}{420}$ of an inch,

Table 1

	At the distance of	
	half a foot	**nine feet**
1. The breadth of the shadow	$\frac{1}{54}$ digit.	$\frac{10}{85} = \frac{2}{17}\ \frac{1}{8\frac{1}{2}}$
2. The breadth between the brightest yellow of the innermost fasciae on either side the shadow	$\frac{1}{40}\ \frac{1}{39}$	$\frac{10}{67}\ \frac{7}{50}\ \frac{3}{20}$
3. The breadth between the brightest parts of the middlemost fasciae on either side the shadow	$\frac{1}{23\frac{1}{3}}\ \frac{1}{23}$	$\frac{4}{17}$
4. The breadth between the brightest parts of the outmost fasciae on either side the shadow	$\frac{1}{18}$	$\frac{10}{33\frac{1}{3}} = \frac{3}{10}$
5. The breadth of the luminous part (green yellow and red) of the first fascia	$\frac{1}{180}$	$\frac{1}{35}$
6. The distance between the luminous parts of the first & 2d fascia	$\frac{1}{240}$	$\frac{1}{42}$
7. The breadth of the luminous part of the 2d fascia	$\frac{1}{260}$	$\frac{1}{52}$
8. The distance between the luminous parts of the second & 3d fascia	$\frac{1}{360}$	$\frac{1}{63}$
9. The breadth of the luminous part of the 3d fascia	$\frac{1}{420}$	$\frac{1}{80}$

as the table seems to imply. By letting the fringes fall obliquely on his ruler or scale, which was divided into $\frac{1}{16}$ of an inch, the observed magnitudes were actually 12 times larger than the values in the table, that is, he was able to observe an interval of about half of $\frac{1}{16}$ of an inch.[8] The multiple values for the entries in lines 2 and 3 indicate Newton's inability to distinguish between them in this early draft. The differences vary from 0.75–7%, but even in the table in the published *Opticks* (see Table 3) Newton still had multiple values. Since the largest difference there is 2.5%, this can serve as a reasonable estimate of the smallest difference that Newton could distinguish by his measurements. Newton "gathered" from these measurements that:[9]

> the rays in the most luminous part of the first fascia passed by the hair at the distance of $0.00767 = \frac{1}{130}$ part of an inch & in their passage by it were bent & turned outwards from the shadow so as to contein with the line produced in which they came from the hole to the hair, an angle of min.

He also reported values for the rays of the next two fringes and for the edge of the shadow, and in each case he left a space for the angle to be inserted. His use of the phrase "the rays in the most luminous part of the first fascia passed by," showed that Newton assumed the identity of the paths of the rays and the fringes. If this is not apparent, it will become perfectly so from his calculation of these distances and angles. Although Newton did not explain how he determined these values, there are a number of worksheets in his papers containing his actual measurements and calculations.

I will first present my reconstruction of Newton's model of diffraction, and then show how he actually used this model in his calculations. Newton assumed (Fig. 3) that the incident rays arriving from point H at the center of the hole are bent away from the hair at a point a short distance $\frac{f}{2}$ away from the center of the hair. According to this model, the rays then proceed in a straight line and depict a fringe of a given order. If the diffracted rays on each side of the hair are projected backwards, they intersect at a point E on the axis of symmetry. Although Newton believed that the rays were gradually deflected as they passed the hair and followed a curved path, for the purpose of measurement and calculation, he assumed the deflection to occur at a point, an assumption that he also made in his work on refraction, where he likewise held that the paths are curved. Calculating the distance f for each fringe (Fig. 4) is a matter of simple geometry. If we let c be the distance between the middle of the fringes on each side of the shadow of the hair (as observed on a screen

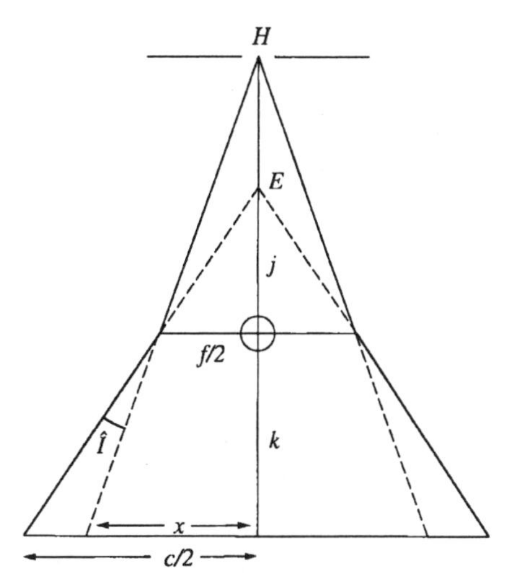

Fig. 3 Newton's model for diffraction by a hair that assumes the paths of the rays and fringes coincide and propagate in straight lines.

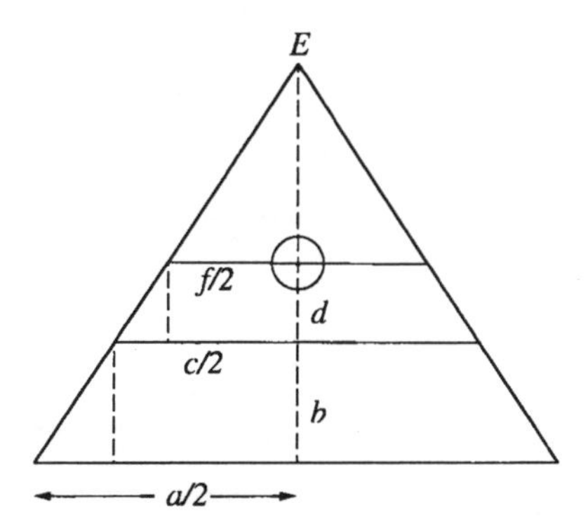

Fig. 4 Determining the points of inflection according to Newton's linear-propagation model for diffraction by a hair.

at a distance d of $\frac{1}{2}$ foot from the hair) and let a be the distance between the middle of the fringes when the screen is nine feet from the hair, then b, the distance of the first screen to the second will be $8\frac{1}{2}$ feet. From similar triangles, we have,

$$\frac{\frac{1}{2}(c-f)}{\frac{1}{2}(a-c)} = \frac{d}{b}$$

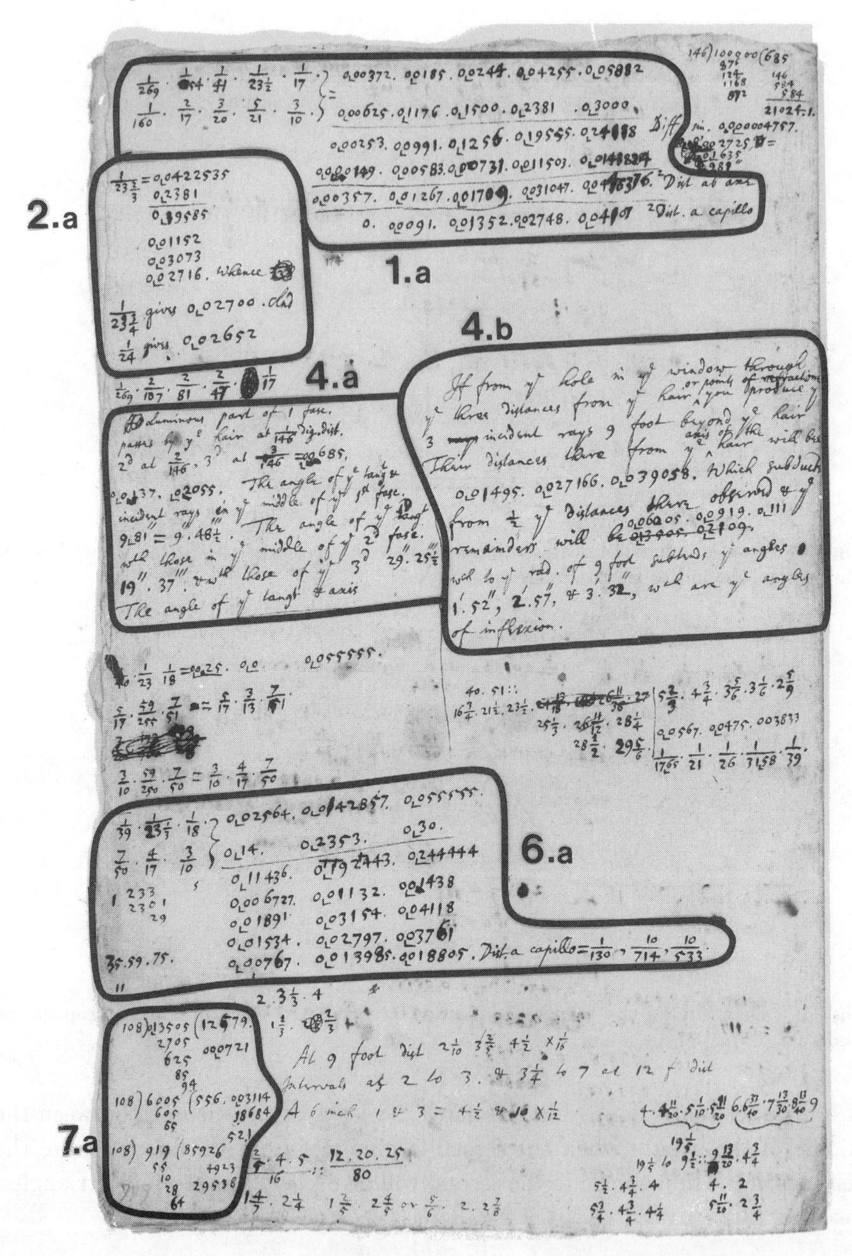

Fig. 5 One of Newton's worksheets on diffraction; Add. 3970, f. 646r. (By permission of the Syndics of Cambridge University Library.)

or

$$f = c - \frac{d}{b}(a - c)\,. \tag{1}$$

Now let h be the diameter of the hair and s the distance of the point of inflection from the surface of the hair, so that

$$s = \frac{f - h}{2}\,. \tag{2}$$

At the beginning of his investigation, Newton determined the diameter of the hair to be $\frac{1}{280}$ of an inch, and he never altered that value. He did not explain how he made that measurement.[10]

Newton left worksheets on diffraction filled with calculations and sketches based on his observations with a hair and knife edges.[11] On one of them, f. 646r, a series of numbers (6.a in Fig. 5) turned out to be his calculation of the inflection distance from the hair as related in the draft on f. 378v that I recently quoted, and we shall see that he used the proportion that I just derived to determine f and s as defined by Eqs. (1) and (2). I have transcribed in Table 2, Newton's sequence of numbers and added the headings and the right-hand column, which explains the steps of the calculation:

These are precisely the numbers that are found in Newton's text except for rounding off. His calculation of the inflection distance was therefore based on

Table 2

Distance between innermost fringes	Distance between middle fringes	Distance between outermost fringes	
$\frac{1}{39} = 0.02564$	$\frac{1}{23\frac{1}{3}} = 0.042857$	$\frac{1}{18} = 0.055555$	c
$\frac{7}{50} = 0.14$	$\frac{4}{17} = 0.2353$	$\frac{3}{10} = 0.30$	a
0.11436	0.192443	0.244444	$a - c$
0.006727	0.01132	0.01438	$\frac{d}{b}\left[= \frac{6\text{ in}}{8.5\text{ ft}} = \frac{1}{17}\right](a - c)$
0.01891	0.03154	0.04118	$f = c - \frac{d}{b}(a - c)$
0.01534	0.02797	0.03761	$f - h\left[= \frac{1}{280} = 0.00357\right]$
0.00767	0.013985	0.018805	$\frac{f - h}{2} = s$
$\frac{1}{130}$	$\frac{10}{714}$	$\frac{10}{533}$	Distance from hair

the assumption that the rays and fringes propagate along identical rectilinear paths. If any ray is intercepted anywhere along its path, it always depicts a fringe of the same order, that is, the rays do not cross or intersect one another.

In further confirmation of my reconstruction, we should recall that Newton had left gaps in the same text to insert his calculation of the "angle of inflection," which he defined as the line between the incident ray produced and the inflected ray. From his worksheets, it is evident that he used the same linear propagation model as in the calculation of the inflection distances. To determine the angle of inflection, he first calculated the distance x from the axis where the extension of the incident ray intersected the screen (Fig. 3). From similar triangles, we find

$$\frac{\frac{1}{2}([f-h]+h)}{j} = \frac{x}{j+k}$$

where j and k, the distances of the hair from the hole and screen respectively, are known. Since Newton had already calculated the distances f of the inflection point from the hair, x could be readily determined for each fringe. Subtracting these distances from half of the distance between the fringes c, he then calculated the angle of inflection $\hat{I}$ (in radians),

$$\hat{I} = \frac{\frac{c}{2} - x}{k} .$$

I was able to reconstruct Newton's method of calculating the angle of inflection from two entries on his worksheets that are based on different values of the distance between the fringes c than those in Table 2. Another passage on his worksheet (see 4.a in Fig. 5) is unusually informative, for Newton actually explained the nature of the calculation,[12] while an actual calculation on yet another worksheet (Add. 3970, f. 357r) illuminated the procedure. This calculation provided additional support for my reconstruction of Newton's model, but I shall spare you the details. Let me now turn to a physical implication of his model and also his experimental confirmation of it.

In this draft that I have been considering, Newton introduced a new empirical result. Observing the fringes on a screen from their first appearance at about 1/4 of an inch from the hair out to nine feet, he found that they "kept very nearely the same proportion of their bredths & intervalls which they had at their first appearing." For example, the ratio of the distance between the middle of the first fringes to that between the third fringes was found to be as "nine to nineteen & by some of my observations near the hair, as six to thirteen."[13] Since these two ratios differ by about 2.5%, and another one by nearly 7%, this provides us with another estimate of what Newton judged acceptable accuracy in his work on diffraction. From his calculations and comparisons of his measurements with his laws, I have found that is about the range of what Newton considered a "reasonable" or "acceptable" result, though generally the upper-bound for error is somewhat smaller, about 5 or 6%.

Had Newton simply come upon an empirical law, or had he deduced it from his model? I am reasonably confident the latter is the case. In pondering the model, some questions naturally arise about the disposition of the sets of rays that are inflected at different distances from the hair and then proceed to depict the three fringes. Are they parallel, or do they diverge, and if they diverge, do they do so from a common point? The last assumption, together with that of the rectilinear propagation of the rays and fringes, immediately leads to the conclusion that the proportion of the fringes is constant at all distances (see Fig. 6). Newton's identification of the paths of the rays and fringes is thus proving to be quite fruitful.

Newton's commitment to empirical confirmation became apparent from the vanishing of his simple linear law for the distance of the point of inflection from the diffracting edge. He made no statements about any law for these

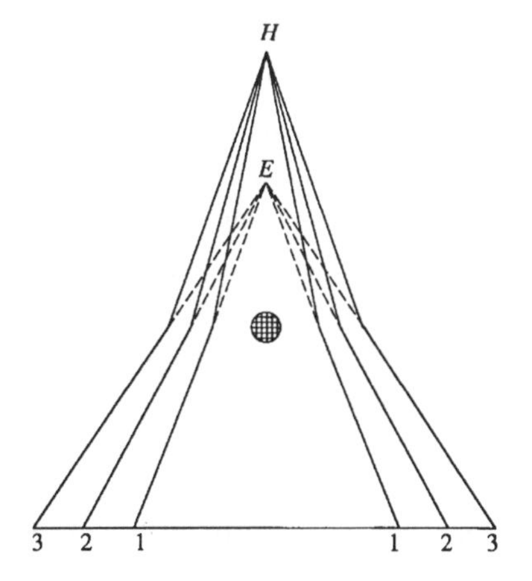

Fig. 6 The three pairs of fringes formed in diffraction by a hair, based on Newton's assumptions that the fringes propagate in the same straight lines as the rays and all the diffracted rays and fringes diverge from a common point.

distances in this draft. From his worksheets, however, it is evident that he had abandoned the old law, because it was not confirmed by his improved measurements. Using his earliest hair measurements, he initially adopted it, for he jotted down "Luminous part of 1 fasc. passes by the hair at $\frac{1}{146}$ dig. dist.[,] 2nd at $\frac{2}{146}$, 3d at $\frac{3}{146} = .00685, 0.0137, .02055$."[14] Testing this linear law against the improved values adopted in this draft, we discover that it differs by 9% for the second fringe and 18% for the third. These are well beyond what I estimated was an acceptable error for Newton in his diffraction experiments, so that we can appreciate why he abandoned it. We should also note how quickly Newton was to generalize his results, both from a limited number of measurements and also from particular cases. He did not, for example, test his model with observations at some distance besides those at $9\frac{1}{2}$ feet, the two that were required to determine the parameters of the model, nor did he try another set of measurements with a hair of a different diameter. Similarly, the linear law was deduced from experiments with a knife edge and then extended to the hair, so that Newton considered it to be a general property of diffraction. He then rejected it for both a hair and knife edges because of its predictive failure with refined measurements with a hair.

Probably not long after Newton composed the draft I have been considering, he wrote out a fair copy of "Observations concerning the inflexions of the rays of light in their passage by the surfaces of bodies at a distance," then Book IV, Part II (Add. 3970, ff. 79–90), in order to conclude the *Opticks.* This part of the *Opticks* was not particularly ambitious. Newton's aim apparently was simply to get a handle on some of the basic features of diffraction and to show that they were consistent with his physical theories, especially on color. Most of the observations were devoted either to diffraction by a hair or various arrangements of knife edges. The observations on the hair were largely a revision of the earlier draft. Newton, however, recorded some new distances, in lines 8 and 10, and eliminated one, the breadth of the third fringe, line 9 in Table 1. From the new table of measurements (Table 3), it can be seen that he changed a number of the values.[15]

The changes for the distance between the middle of the three fringes (lines 2–4 in each table) were not great; they were either nil or on the order of 2–2.5%. For the more delicate measurements in lines 7 and 9 (corresponding to lines 5 and 7 in Table 1), the changes were larger, ranging from 5.5%–11.5%.[16] Just as in the preceding draft, Newton calculated the distances of the inflection point from the hair and also the angle of inflection. On the basis of these more

Table 3

	At the distance of	
	half a foot	**nine feet**
1. The breadth of the shaddow	$\frac{1}{54}$ dig.	$\frac{1}{9}$ dig.
2. The breadth between the middles of the brightest light of the innermost fasciae on either side the shaddow	$\frac{1}{38}$ or $\frac{1}{39}$	$\frac{7}{50}$
3. The breadth between the middles of the brightest light of the middlemost fasciae on either side the shaddow	$\frac{1}{23\frac{1}{2}}$	$\frac{4}{17}$
4. The breadth between the middles of the brightest light of the outmost fasciae on either side the shaddow	$\frac{1}{18}$ or $\frac{1}{18\frac{1}{2}}$	$\frac{3}{10}$
5. The distance between the middles of the brightest light of the first & second fascia	$\frac{1}{120}$	$\frac{1}{21}$
6. The distance between the middles of the brightest light of the second & third fascia.	$\frac{1}{170}$	$\frac{1}{31}$
7. The breadth of the luminous part (green white yellow and red) of the first fascia	$\frac{1}{170}$	$\frac{1}{32}$
8. The distance of the darker space between the first & second fascia	$\frac{1}{240}$	$\frac{1}{45}$
9. The breadth of the luminous part of the second fascia	$\frac{1}{290}$	$\frac{1}{55}$
10. The breadth of the darker space between the 2^{d} & 3^{d} fascia	$\frac{1}{340}$	$\frac{1}{63}$

refined values, he announced new laws for both the distances and the angles of inflection, namely, that "the squares of these distances are in the arithmetical progression of the odd numbers 1, 3, 5, without any sensible errors & the squares of these angles of inflexion are also in the same progression." Thus, the new laws are

$$s^2 \propto 1, 3, 5 \quad \text{and} \quad \hat{I}^2 \propto 1, 3, 5$$

or

$$s \propto \sqrt{1}, \sqrt{3}, \sqrt{5} \quad \text{or} \quad \hat{I} \propto \sqrt{1}, \sqrt{3}, \sqrt{5}\,.$$

This new law differs significantly from the old linear law for the distances that Newton first proposed for diffraction from a knife edge. No measurements survived from this earliest period, so the basis of that law must have been very casual observations. Newton's new measurements agreed with this new law with an error ranging from about 0.5–4%, which according to my assessment was within the limit of his tolerance.[17]

Newton returned to the proportionality of the breadth and intervals of the fringes in the next observation, and introduced a greater number of measurable proportions. The differences between the law and the measurements were mostly around 1.5%, but all remained well within his acceptable range. He added yet another law (written in the margin of the manuscript):

> the breadths of the fasciae seemed to be in the progression of the numbers 1, $\sqrt{\frac{1}{3}}$, $\sqrt{\frac{1}{5}}$, & their intervalls to be in the same progression with them, that is the fasciae & their intervalls together to be in the continuall progression of the number[s] 1, $\sqrt{\frac{1}{2}}$, $\sqrt{\frac{1}{3}}$, $\sqrt{\frac{1}{4}}$, $\sqrt{\frac{1}{5}}$ or there abouts.[18]

This new law agrees remarkably well with the data (all within 1.5%) and fits the data the best of all Newton's diffraction laws. The measurements in the preceding draft, Table 1, do not lead to this law, since they differ from it by about 5–17%. The values in lines 5 and 6 (corresponding to lines 6 and 8 in Table 1) were changed by a factor of two, and the earlier values undoubtedly were caused by some error in reading or conversion. Almost no one, let alone Newton, made experimental errors of that order. Although Newton did fiddle with his measurements (usually by no more than a few percent), I do not believe that his aim here was to get better agreement with his law. Since the law was added to the manuscript after it was written, it appears that he had not yet come upon it when he entered these new values. Moreover, these values, unlike many others in the table, were not subsequently altered in the manuscript. This law does not follow from his model, as the simple proportion does, and was no doubt derived inductively. It is clear, though, that he was searching for some regularity of this sort because of the new measurements in lines 8 and 10.

What do I mean when I say that Newton "tinkered" with his measurements? The words "fiddling" or "tinkering" were suggested to me by the manuscripts where one sees a range of numbers differing from one another by small amounts, from which Newton finally chose one. Newton, it seems, was trying out slightly different values to find which yielded better agreement with his laws or a more evident law. These were all measured values or within the range of his measured values. For example, in 1.a in Fig. 5, Newton carried out his calculation with $23\frac{1}{2}$, but in 2.a we can see him trying out what he would get with $23\frac{2}{3}$, $23\frac{3}{4}$ and 24. He had determined a range for his measurements of the second fringe (as well as the others) as can be seen in row 3 in Tables 1 and 3. Newton seems to have been playing with his numbers within what he judged to be his experimental error in order to get a better fit with his various laws.[19] It is not apparent how he finally decided which value to use. This process of massaging the data (as contemporary scientists call it) may improve agreement with the laws, but it is risky. We shall encounter an example of the risky side in Newton's renunciation of the linear propagation model that he added to Observation 1 during the final revision of the manuscript of the *Opticks*.

Newton, it must be emphasized, never expounded his linear propagation model in this first, and soon to be suppressed, state of the art on diffraction. He said nothing about the way in which he had determined the distances and angles of inflection, s and $\hat{I}$, that appeared in his laws describing diffraction; he simply stated their values. I had to uncover their physical foundation, the linear propagation model from his worksheets and calculations. Newton's methodology compelled him to suppress his model, because it was a hypothesis imposed on the observations and not derived from them. He believed that hypotheses or conjectures could not be mixed with the more certain part of science, the phenomena or principles derived from the phenomena. To do so would compromise the quest for a more certain science.[20] Thus, the linear propagation model lay hidden in the background, underlying his data and guiding his search for the laws governing diffraction.

After treating diffraction by a hair, Newton turned to diffraction by knife blades. We should recall that, in his earliest observations on diffraction, he found that the edge of a single knife produced three fringes, and that when he placed the knives so that they formed a V or wedge (Fig. 1), he found two sets of intersecting fringes (Fig. 2). While he was carrying out his investigation of diffraction by a hair, he also continued his study of diffraction with knife edges. In Observation 8, he carefully repeated the experiment in which two knife edges formed a V. The blades now made an angle of $1° \ 54'$ and were placed ten or 15 feet from the hole, which was $\frac{1}{42}$ of an inch in diameter. First, let me present a brief overview of Newton's description of the fringes formed by the inclined blades, in which I shall identify the paths of the fringes and rays, just as he did. The fringes are observed on a screen (a white ruler) that is gradually moved further away from the two knives. As the incident rays pass each edge, they are bent away from it and depict three fringes parallel to that edge; as they move further away from the two knives, the two sets of fringes/rays cross one another. Moreover, the rays that pass through the V closer to the intersection of the blades, that is, where the edges are closer to one another, cross one another nearer to the two knives than those higher up. Thus, if the screens are placed further from the knives than they were in Newton's figure, the intersections of the fringes would shift to the left in his figure.

Let us now turn to Newton's own account. When the light fell on a white ruler about an inch beyond the knife blades, he

> saw the fasciae made by the two edges run along the edges of the shadows [of the knives] in lines parallel to those edges without growing sensibly broader, till they met in angles equal to the angle of the edges, & where they met & joyned they ended without crossing one another. But if the ruler was held at a much greater distance from the paper they became something broader as they approached one another & after they met they crossed & then became much broader then before.

From these observations he concluded:

> Whence I gather that the distances at which the fasciae pass by the knives are not increased nor altered by the approach of the knives but the angles in which the rays are there bent are much increased by that approach. And that the knife which is nearest any ray determins which way the ray shall be bent & the other knife increases the bent.[21]

The first conclusion follows from the fact that the fringes are parallel to the knife edges. The second conclusion, which also satisfies observation, is physically odd, for it implies that the edge closer to the rays bends them away, while the opposite one "attracts" them.

Newton reported in Observation 9 that at $\frac{1}{3}$ of an inch from the knives, he found that the shadows between the first two fringes crossed one another $\frac{1}{5}$ of an inch above the intersection of the blades. He readily calculated that the distance between the blades was $\frac{1}{160}$ of an inch, so that the "dark intervalls of the first & second fascia meeting in the middle of this distance are in their passage beetween the knives distant from their edges $\frac{1^{\text{th}}}{320}$ part of an inch." Now he invoked the "law" of Observation 3 — which was derived from measurements on the *hair* and assumed identical rectilinear paths of propagation for the rays and fringes — that the "the bright fascias & their intervalls pass by the edge of the knife at distances which are in the progression of the numbers $\sqrt{1}.\sqrt{2}.\sqrt{3}$ & c."[22] With this law, he calculated that the three bright fringes passed by the edge of the *knife* at $\frac{1}{452}$, $\frac{1}{261}$ and $\frac{1}{202}$ of an inch. This implies that the light from the second and third fringes comes from the half of the light closer to the opposite edge, which seems to contradict his observation that the fringes are produced by the light along the closer edge. Another hidden implication — and this would be the fatal one — is that if he had made his measurement at any other position, he would have found that the light producing each of the fringes passes the edges at different distances. Newton, however, was for the time being content with these results, which are included in the first completed state of the *Opticks*.

Newton became dissatisfied with his newly completed Book IV some time between the fall of 1691 and February 1692 and dismembered it. He revised Part I on the colors of thick plates and made it Part IV of the preceding book; he also devised the theory of fits and added a number of propositions on fits at the end of Book III, Part III. He was, however, sufficiently dissatisfied with Part II on diffraction that he removed it from the manuscript of the *Opticks* without adding a revised replacement. By comparing the suppressed version with the published one that Newton wrote around 1703, it is apparent that a new set of observations that he had made with the inclined knives had undermined his model for diffraction and forced him to recognize that the paths of the rays and the fringes were distinct. The experiment dates from the period around February 1692, when he removed the just completed part on diffraction from the *Opticks*, for it is written on the top of a single folio that is immediately followed by an account of halos observed in June 1692. Newton decided to measure the distance between the knife edges when the intersection of the first dark lines fell on a white paper placed at different distances from the knives (Table 4).[23]

These measurements showed that the light forming the *same* fringe or dark interval, when observed at various distances from the knives, came from different distances from the edges and was deflected at different angles. When the paper, for example, was $1\frac{1}{2}$ inches from the knives, the light passed at most 0.006 inch from each edge, whereas at 131 inches, it was at most 0.043 inch away. Moreover, the angle through which the rays were diffracted (assuming that they were diffracted at the edges of the knives) varied from nearly 30′

Table 4

Distances of the Paper from the Knives in Inches	Distances Between the edges of the Knives in millesimal parts of an Inch
$1\frac{1}{2}$	0'012
$3\frac{1}{3}$	0'020
$8\frac{3}{5}$	0'034
32	0'057
96	0'081
131	0'087

at the former to just 2′ at the latter. Since light propagates in straight lines after it has passed through the blades and been deflected, it cannot be the same light that formed the fringes at different places, that is, the fringes do not propagate rectilinearly. Newton's record of this experiment from 1692 contains no comments on its implication. When he later related this result in the *Opticks*, he tersely noted:

> And hence I gather that the Light which makes the fringes upon the Paper is not the same Light at all distances of the Paper from the Knives, but when the Paper is held near the Knives, the fringes are made by Light which passes by the edges of the Knives at a less distance, and is more bent than when the Paper is held at a greater distance from the Knives.[24]

As important as this conclusion is, it nonetheless renders much of Newton's work on diffraction invalid, since he had carried out many of his investigations and derived many of his conclusions based on the assumption that the rays and fringes followed identical rectilinear paths.[25] Newton no doubt recognized the implications of this experiment as soon as he performed it. He removed the part on diffraction from Book IV of the *Opticks*, while he added the theory of fits and shifted its revised companion on the colors of thick plates to Book III.

About two years later, when David Gregory visited Newton in Cambridge in May 1694 and was allowed to study the manuscript of the *Opticks*, the first three books were finished, but Newton had done nothing about restoring the part on diffraction. Gregory noted that "I saw *Three Books of Opticks*.... The fourth, about what happens to rays in passing near the corners of bodies, is not yet complete. Nonetheless, the first three make a most complete work."[26] While Newton had demonstrated the untenability of the model he had been using, there was no intrinsic reason that his work on diffraction had to come to a halt. Earlier in his career, he had rejected his own initial approaches and then gone on to develop new ones. Newton did intend to start over, as he related in the conclusion added to the published *Opticks*:

> When I made the foregoing Observations, I designed to repeat most of them with more care and exactness, and to make some new ones for determining the manner how the rays of Light are bent in their passage by Bodies for making the fringes of Colours with the dark lines between them. But I was then interrupted, and cannot now think of taking these things into further consideration.[27]

Newton was now near the end of his active scientific career, and, despite his intentions he never returned to experimental research on diffraction. In April 1695, he told John Wallis, who was trying to convince him to publish the *Opticks*, that the last part of it was still incomplete, and a year later he had moved to London to assume the post of Warden of the Mint.[28]

When Newton decided to publish the *Opticks* in late 1702, he chose to revise the final book on diffraction rather than to eliminate it altogether.[29] Most of the revision entailed removing passages that were based on the linear propagation model, for example, the calculations of the distance of the points of inflection from the diffracting edge and the laws governing them. In Observation 9, he added the measurements that had invalidated that model.[30] Since Newton developed his model of the linear propagation of the fringes in his observations on the hair, it was only fitting that he renounced that model in a major addition that he made to Observation 1. He introduced an illustration of diffraction by a hair (Fig. 7) that clearly showed the rays crossing one another, so that the paths of the fringes were not the same as that of the rays. He explained:

> the Hair acts upon the rays of Light at a good distance in their passing by it. But the action is strongest on the rays which pass by at least distances, and grows weaker and weaker accordingly as the rays pass by at distances greater and greater, as is represented in the Scheme: For thence it comes to pass, that the shadow of the Hair is much broader in proportion to the distance of the Paper from the Hair, when the Paper is nearer the Hair than when it is at a great distance from it.[31]

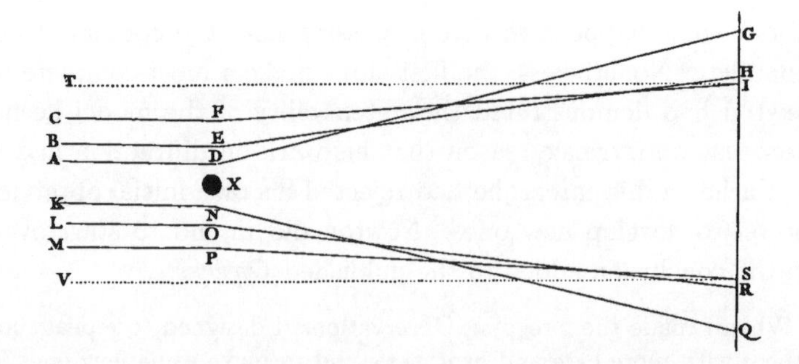

Fig. 7 Diffraction by a hair; from the *Opticks*, Book III, Fig. 1.

Newton writes as if the nonlinear propagation of the shadow, and hence, the crossing of the rays is obvious and follows directly from observation. On the contrary, his data do not demonstrate that the rays must cross or that the linear propagation model must be abandoned. Newton, I believe I have shown, came upon the crossing of the rays during the course of his observations with knife edges and not the hair. That the data do not support Newton's claim further supports this interpretation. When Newton revised the part on diffraction for publication and added this new paragraph to the manuscript of Observation 1, he did not present any new experimental evidence. Rather, he let the old observations stand. In Observation 1, he reported three measurements of the breadth of the shadow — at four inches, and two and ten feet — but he had also included two measurements of the shadow along with his measurements of the fringes in Observation 3 (see line 1 in Table 3). Newton justified the crossing of the diffracted rays (which implies the curvilinear propagation of the fringes) by arguing that the shadow is disproportionally broader when it is measured closer to the hair than further away. The measurements in Observation 1 do support this claim, but when those in Observation 3 are added, it becomes a bit ambiguous.[32] Nonetheless, if the disproportion is accepted as valid, it may still be readily accounted for by his original linear propagation model, where the diffracted rays diverge from a common point of intersection (Fig. 6). When I compared Newton's measurements of the breadth of the shadow at various distances from the hair with the values calculated according to his original model, they differed only by an undetectable fraction of a percent to 2.5%.[33] Thus, the rays need not cross.

If Newton's claim that the rays cross one another because the shadow is disproportionally broader when it is observed closer to the hair than further away is correct (and it is), it is natural to ask why his data does not support that claim. In the first place, the discovery of the crossing of the rays was made in another experiment with knife blades and then announced in Observation 1 at the beginning of his account of diffraction. In the second place, it may be strongly suspected that Newton had earlier fiddled with his observations so that they would agree better with the linear propagation model that he was then using.[34] The shadow has to propagate according to the same laws as the rays, since its boundary is defined by the rays (namely, by the innermost edges of the first fringe). Newton was certainly aware of this, and in a number of his calculations for the linear propagation model, similar to the one that I presented in Table 2, he included the shadow.[35] Since the shadow is not at all well-defined (the intensity of the light decreases gradually from the first fringe), it is difficult to measure. Consequently, Newton had some leeway for exercising his judgment in measuring the shadow's breadth and forging agreement with his linear propagation model. A reasonable conjecture, then, to explain the incongruence between Newton's data and his interpretation of it is that after he performed his experiment with the knife blades and found that the rays cross one another, he then went back to Observation 1 and re-interpreted his original "massaged" data.

If my account of Newton's investigation of diffraction for the *Opticks* is correct, then the traditional explanation for the delay of its publication, namely, that he was waiting for Hooke to pass from the scene (he died in March 1703) is not altogether sufficient. For much of the period between the purported completion of the *Opticks* and its publication in 1704, there was no *Opticks* to publish. It was incomplete. I suspect that at least until 1697, and perhaps for another few years, Newton intended to return to his experiments on diffraction. Until he fully accepted that he would not take them up again and would instead edit and revise the original completed part, there was no *Opticks* to delay. There is no doubt that Newton's wish to avoid controversy — especially with his nemesis, Hooke — was as much a factor in his holding back the publication of the *Opticks* as its incompleteness, as he told John Wallis in 1695, but to place the entire burden for its long delay on Hooke is unfair.[36]

One might imagine, incorrectly, that Newton's failure to develop a physical explanation for diffraction might be partly responsible for his postponing publication. I do not intend to examine Newton's search for a physical explanation of diffraction, but I can indicate the problems that continually frustrated him, namely, explaining what caused dark bands to appear between the bright ones and what happened to the light from those dark bands. Explaining the deflection of light into or away from the shadow was in principle a simple matter for Newton. Before the *Principia*, he invoked variations in the density of the aether, which would cause the light rays to change direction or bend, and in the *Principia* and afterwards he replaced the aether by attractive and repulsive forces. To explain the alternating bright and dark fringes, it was natural for Newton to attempt to relate them to the similar alternating bright and dark rings he had observed in thin films and the vibrations or fits that he used to explain them. He suggested this mechanism in 1675 in his first discussion of diffraction in his "An Hypothesis Explaining the Properties of Light," and again in about 1690 in drafts of the "Fourth Book," which are early sketches for the conclusion of the *Opticks*.[37] He soon abandoned this idea, presumably because he never had any evidence that periodicity was involved in diffraction, and also because once he had formally specified the properties of the fits to explain the colors of thin and thick plates, they were incompatible with diffraction. For example, the fits acted in the direction of propagation, whereas for diffraction he needed a cause that would act transversely. Newton simply could not explain the cause of the dark fringes and what happened to that light. For a while, he thought that the rays from the dark bands were inflected into the shadow "so as faintly to illuminate all the dark space behind the body," but he had to abandon that idea too.[38] In the *Opticks*, all he could do was lamely ask whether the rays are not "in passing by the edges and sides of Bodies, bent several times backwards and forwards, with a motion like that of an Eel? And do not the three fringes of coloured Light above-mentioned, arise from three such bendings?"[39] Nonetheless, that Newton did not possess a physical explanation for the cause of diffraction did not hinder him from pursuing his experimental research on diffraction or publishing it, for he never required a causal explanation of the phenomena whose laws he had described.

Newton's investigation of diffraction may seem to some to be too slipshod to be worthy of the great man, as he fiddled with his numbers, derived imaginary laws, and developed an untenable physical model. In fact, I find that these are characteristics of all of his optical research. Newton — in a way that I do not yet fully comprehend — played off his measurements against his mathematical descriptions, which allowed him both to change his measurements and to revise his laws. He did not change them at will to get perfect agreement, but controlled them by an awareness of his experimental error and by additional experiments. The only unusual feature of Newton's investigation of diffraction is that he did not unrelentingly pursue it until he had control over the phenomenon as he did earlier with chromatic dispersion and the colors of thin plates, and — nearly simultaneous with his work on diffraction — the colors of thick plates. Far from being slipshod, Newton's critical standards and rigorous methodology are in fact what led him to reject and withdraw his own initial investigation of diffraction. By continually striving to quantify phenomena, developing laws that could be tested, devising new experiments to extend and test his results, and setting realistic measures of experimental error, Newton controlled his fertile imagination by experiment, which in turn, stimulated his imagination. By peering in Newton's "workshop" to examine his notes, drafts, data, and worksheets, we can begin to grasp how Newton actually worked at his science.

Acknowledgment

I thank Michael Nauenberg for a stimulating exchange on the accuracy of Newton's diffraction measurements following the delivery of my paper. He has since developed these comments into the paper "Comparison of Newton's diffraction measurements with the theory of Fresnel" included in this volume.

Notes and References

1. Until shortly before publication, the books in the manuscript of the *Opticks* were numbered differently from those of the published edition. Parts I and II of Book I of the published *Opticks* were Books I and II in the manuscript (which is now in Cambridge University Library), so that the manuscript had four books rather than the three of the published edition. I use the published numbering except in some references to the manuscript. See Alan E. Shapiro, *Fits, Passions, and Paroxysms: Physics, Method, and Chemistry and Newton's Theories of Colored Bodies and Fits of Easy Reflection* (Cambridge, Cambridge University Press, 1993), pp. 139, 149–150.
2. Grimaldi, *Physico-mathesis de lumine, coloribus, et iride, aliisque adnexis libri duo* (1665; reprint, Bologna: Arnaldo Forni, 1963); Fabri, *Dialogi physici quorum primus est de lumine ....* (Lyon, 1669). For Hooke's lecture, see Thomas Birch (ed.) *The History of the Royal Society of London, for Improving of Natural Knowledge, From Its First Rise*, four volumes (1756–1757; reprint, Brussels: Culture et Civilisation, 1968), Vol. 3, pp. 194–195; and Hooke, *The Posthumous Works of Robert Hooke, M.D. S.R.S., Geom. Prof. Gresh. &c. Containing His Cutlerian Lectures, and*

Other Discourses, Read at the Meetings of the Illustrious Royal Society (ed.) Richard Waller (1705; reprint, New York: Johnson Reprint, 1969), pp. 186–190. For accounts of Newton's work on diffraction see Roger H. Stuewer, "A critical analysis of Newton's work on diffraction," *Isis* **61** (1970): 188–205; and A. Rupert Hall, "Beyond the fringe: diffraction as seen by Grimaldi, Fabri, Hooke and Newton," *Notes and Records of the Royal Society of London* **44** (1990): 13–23.

3. Shapiro, "Beyond the dating game: watermark clusters and the composition of Newton's *Opticks*," in *The Investigation of Difficult Things: Essays on Newton and the History of the Exact Sciences in Honour of D.T. Whiteside* (eds.) P.M. Harman and Alan E. Shapiro (Cambridge, Cambridge University Press, 1992), pp. 181–227; and *Fits, Passions, and Paroxysms*, pp. 138–150.
4. See Shapiro, *Fits, Passions, and Paroxysms*, Chap. 4.1.
5. Cambridge University Library, MS Add. 3970, f. 372v; henceforth cited as Add. 3970. Newton also included this law in two drafts of the "Fourth Book," Proposition 2, ff. 335r, 338v; on the "Fourth Book," see Shapiro, *Fits, Passions, and Paroxysms*, pp. 141–143. For clarity and simplicity in my quotations from Newton's manuscripts, I will not indicate the many textual changes, unless they are of significance for this study. In the *Opticks*, Newton concluded that the fringes formed hyperbolas; see Newton, *Opticks: Or, a Treatise of the Reflexions, Refractions, Inflexions and Colours of Light* (1704; reprint, Brussels: Culture et Civilisation, 1966), Book III, Observation 10; and for modern theoretical and experimental accounts of the experiments with the inclined knives as presented in the *Opticks*, see M.P. Silverman and Wayne Strange, "The Newton two-knife experiment: intricacies of wedge diffraction," *American Journal of Physics* **64** (1996): 773–787; and Nauenberg, "Comparison of Newton's diffraction measurements" in this volume.
6. Add. 3970, Proposition 3, f. 377r.
7. Add. 3970, f. 378r,v. I have added the line numbers in the left-hand column and edited the manuscript slightly. Newton subsequently crossed out the observation in line 9. There is a (faint) bright fringe at the center of the shadow of the hair that Grimaldi had observed but which Newton failed to detect; see Stuewer, "A critical analysis."
8. The term "fascia" was used by both Grimaldi, *Physico-mathesis de lumine*, p. 3, and Fabri, *Dialogi physici*, p. 11. It is evident from a sheet on which

Newton recorded his measurements, that his scale was divided into $\frac{1}{16}$ of an inch; Add. 3970, f. 373r. Since the intervals were increased by a factor of 12, Newton had inclined his ruler at an angle of 5° to the beam, that is, nearly parallel to it.

9. Add. 3970, f. 378v.
10. Peter Spargo has proposed a method, which was available to Newton, by which he could have measured the diameter of his hair; " 'Y^{e} exactest measure I could make ...': A possible explanation of Newton's determination of the thickness of a hair," *Transactions of the Royal Society of South Africa*, **50** (1995): 165–168. Although Newton's measurement of the size of his hair is remarkable for its date, it is almost certainly much too large a value.
11. This worksheet, Add. 3970, ff. 645–646, is written on a folded sheet that has an undated draft of a letter to the Commissioners of Taxes on one side. Newton used all four sides for his calculations. The letter is tentatively dated to the early 1670s in Newton, *The Correspondence of Isaac Newton* (eds.) H.W. Turnbull, J.F. Scott, A. Rupert Hall and Laura Tilling, seven volumes (Cambridge, Cambridge University Press, 1959–1977), Vol. 7, p. 371, and to about 1675 by Richard S. Westfall, *Never at Rest: A Biography of Isaac Newton* (Cambridge, Cambridge University Press, 1980), p. 209, n. 89. This dating must be revised. The diffraction data and calculations as well as the watermark argue for the early 1690s. The watermark that I designate GLC was used in other parts of the manuscript of the *Opticks*, for most of the drafts on diffraction for the *Opticks*, and the suppressed part on diffraction; see Shapiro, "Beyond the dating game," p. 202, Table 1.
12. "If from the hole in the window through the three distances from the hair or points of refraction you produce the 3 incident rays 9 foot beyond the hair. Their distances there from the axis of the hair will bee 0.01495. 0.027166. 0.039058. Which subducte from $\frac{1}{2}$ the distances there observed & the remainders will be 0.06005. 0.0919. 0.111 which to the rad[ius] of 9 foot subtends the angles $1'.52''$, $2'.57''$ & $3'.32''$, which are the angles of inflexion" (Add. 3970, f. 646r). In all of Newton's calculations, $j = 12$ ft., $k = 9$ ft. and $h = \frac{1}{280}$ in. He used the values for c and f that are in the calculation on the top of the sheet at 1.a and another calculation is in the bottom right-hand corner at 7.a.
13. Add. 3970, ff. 378v, 328r.

14. *Ibid.*, f. 646r; see 4.a. in Fig. 5.
15. *Ibid.*, f. 81r. I have added the line numbers in the left-hand column.
16. From Nauenberg's comparison of Newton's measurements of the distance between the three fringes with that calculated according to modern theory, it can be seen that Newton's measurement at $\frac{1}{2}$ foot differ from the calculated values by 1.5 to 4.5%, while at nine feet they range from 1.7 to 2.4%; "Comparison of Newton's diffraction measurements." This shows that Newton had achieved a high degree of accuracy with his measurements.
17. *Ibid.*, f. 82r. Newton had confidently written that "if they could be measured more accurately it's without any sensible errors," before he decided to delete it and let the reader make his own assessment.
18. *Ibid.*, f. 83r.
19. Another example of Newton trying out a set of values can be found in his fringe measurements in Add. 3970, ff. 377v, 378r (Table 1). On the sheet facing the table of his draft, he had four more sets of values for the key first four lines, some of which differed by 7 or 8%; he published the third of these. We should recall that these measurements stand up quite well against the predictions of modern theory (see note 16).
20. On Newton's quest to construct a hypothesis-free science and how it affected the formulation of his theory of fits, see Shapiro, *Fits, Passions, and Paroxysms*, Chaps. 1.2 and 4.3.
21. Observation 8, *ibid.*, f. 87r–88r. This passage is essentially unchanged (except for the switch from "fasciae" to "fringes") in the published *Opticks*, Book III, Observation 7, p. 125.
22. Add. 3970, f. 88r.
23. The values in the *Opticks* (Book III, Observation 9, p. 127) scarcely differ from the original measurements on Add. 3970, f. 334r, and I give those from the *Opticks.* The observation of the halo was subsequently incorporated in Book II, Part IV, Observation 13. The watermark on f. 334 is consistent with the early 1690s, since it is the same as that used for the definitions that begin Book I. I have designated this watermark "None" in the two tables in "Beyond the dating game," pp. 202–203.
24. Newton, *Opticks*, Book III, Observation 9, p. 127. There is a draft from around 1703 of this passage in Add. 3970, f. 477v.
25. David Brewster judged this to be one of Newton's "two new and remarkable results" on diffraction; *Memoirs of the Life, Writings, and Discoveries*

of Sir Isaac Newton, two volumes (1855; reprint, New York: Johnson Reprint, 1965), Vol. 1, p. 200.

26. Gregory, *Notae in Newtoni Principia*, Royal Society, MS 210, insert between pp. 55, 56; my translation; for the Latin see Shapiro, *Fits, Passions, and Paroxysms*, p. 148.
27. Newton, *Opticks*, Book III, p. 132.
28. Wallis to Newton, 30 April 1695, Newton, *Correspondence*, Vol. 4, p. 117.
29. According to a memorandum by David Gregory on 15 November 1702: "He [Newton] promised Mr. Robarts, Mr. Fatio, Capt. Hally & me to publish his Quadratures, his treatise of Light, and his treatise of the Curves of the 2^d Genre;" W.G. Hiscock (ed.) *David Gregory, Isaac Newton and Their Circle: Extracts from David Gregory's Memoranda 1677–1708* (Oxford: Printed for the Editor, 1937), p. 14.
30. In the Advertisement to the *Opticks*, pp. [iii–iv], Newton felt it necessary to warn the reader that "The Subject of the Third Book I have also left imperfect, not having tried all the Experiments which I intended when I was about these Matters, nor repeated some of those which I did try, until I had satisfied myself about all their Circumstances." During revision he added the concluding Observation 11, which describes diffraction in monochromatic light, but this observation goes back to one of his earliest drafts on diffraction from the early 1690s, Add. 3970, ff. 324–325. I cannot now explain why Newton omitted it from the first completed state of the part on diffraction, Book I, Part II.
31. Newton, *Opticks*, Book III, Observation 1, pp. 115–116.
32. The breadth of the shadow at distances of four inches, six inches, nine feet and ten feet from the hair were $\frac{1}{60}$, $\frac{1}{54}$, $\frac{1}{28}$, $\frac{1}{9}$ and $\frac{1}{8}$ of an inch, respectively. The ratios of the breadth to the distance are then $\frac{1}{240} > \frac{1}{324} > \frac{1}{672} > \frac{1}{972} < \frac{1}{960}$. Since the difference in the last two ratios is not great, one could attribute the discrepancy to experimental error.
33. Using Eq. (1), I calculated from Newton's measurements the distance from the center of the hair, $\frac{f}{2}$, at which the rays at the edge of the shadow are inflected. It was then a simple matter to compare the breadth of the shadow predicted by the linear propagation model with the observed values.
34. Michael Nauenberg first called my attention to the fact that Newton's observations were inconsistent with Fresnel's theory. I had already recognized that Newton's data were inconsistent with the claim that the rays must cross.

35. See, for example 1.a in Fig. 5, where the first column treats the calculated geometrical shadow, the second the measured "physical" shadow, and the next three the three fringes; see also Add. 3970, f. 646v.
36. Wallis to Newton, 30 April 1695, Newton, *Correspondence*, Vol. 4, pp. 116–117; Newton's letter to Wallis is lost, but his reasons for not publishing the *Opticks* can be readily reconstructed from Wallis's counter-arguments.
37. "Hypothesis," Newton, *Correspondence*, Vol. 1, pp. 384–385. "Fourth Book," Proposition 13, Add. 3970, f. 335v; and Proposition 10, ff. 337r,v.
38. "Fourth Book," Proposition 3, Add. 3970, f. 338v.
39. Newton, *Opticks*, Query 3, p. 133.

Michael Nauenberg

Comparison of Newton's Diffraction Measurements With the Theory of Fresnel

MICHAEL NAUENBERG

Department of Physics, University of California, Santa Cruz CA 95064, U.S.A.

Abstract

In the *Opticks*, Newton reported detailed measurements on diffraction fringes from various "slender substances" placed in a beam of sunlight. The accuracy of these measurements is examined by comparing them with predictions obtained from analytic and numerical solutions of the Fresnel theory for diffraction. Newton's diffraction experiments with a hair are discussed, and it is shown that the fringe separation is independent of the width of the hair. Uncertainties in this width were previously thought to be a source of error in attempting to determine the precision of Newton's observations. The remarkable fringe pattern for Newton's wedge-shaped slit drawn in the *Opticks* is discussed and shown to be invariant with the distance of the observer from the slit. It is argued that the scaling of this pattern with distance was an unsurmountable stumbling block for Newton's corpuscular theories of diffraction.

At this symposium, Alan Shapiro gave a lecture on Newton's experimental investigation and early theories of diffraction based on Newton's unpublished optical manuscripts.[1] This talk suggested to me the need to examine the accuracy of Newton's measurements, which turn out to be crucial in the development of Newton's ideas about the nature of diffraction. The purpose of this note is to compare the diffraction measurements which Newton reported in the *Opticks* with the predictions of Fresnel's modern theory of diffraction. With the exception of a few cases, it will be shown that the agreement is remarkably good, particularly considering the difficulty of Newton's pioneering experiments. For diffracting objects, Newton used a hair and a slit of variable width consisting of two knives with their edges tilted at a very small angle.[1] Recently, these experiments have been repeated,[2,3] but a quantitative comparison between Newton's measurements and the predictions of Fresnel's wave theory has not been made before. In his *Opticks*,[4] Newton described his experimental arrangement as follow (see also A. Shapiro; Chapter 4)

"I made in a piece of Lead a small Hole with a Pin, whose breadth was the 42nd part of an Inch... Through this Hole I let into my darken'd

> Chamber a beam of the Sun's Light and found that the Shadows of hairs, Tred, Pins, Straws and such like slender substances placed in this beam of Light, were considerably broader than they ought to be, if the Rays of Light passed on by these Bodies in right Lines ... The Shadows of all Bodies (Metals, Stones Glass, Wood, Horn, Ice, & c.) in this Light were border'd with three Parallel Fringes or Bands of colour'd Light ... "

Newton made detailed measurements of the spacing of these fringes which he observed on a screen at various distances from the diffracting object described in a series of ten observations in the *Opticks.* However, he failed to provide a consistent theory as Shapiro discusses in detail in this volume.[1] Almost a hundred years after the publication of the *Opticks*, Thomas Young showed that the diffraction fringes could be explained by the interference of light waves propagating directly from the source with light waves coming from the edges of the object.[6,7] In fact, Young was motivated by Newton's diffraction measurements which he applied directly to verify his theory as well as some further measurements of his own. Theoretical ideas similar to Young's were developed 15 years later by Augustin Fresnel who gave a more complete account of diffraction by combining a modified form of Huygens principle for the propagation of a light front with the interference principle of the wave theory.[8] A critical factor in the Fresnel theory is the dimensionless parameter $c = \pi a^2(1 + z/L)/(\lambda z)$, where a is the width of the diffracting object or slit, L is the distance of the source, z is the distance of the screen from the hair, and λ is the wavelength in the case of incident monochromatic light. For values of $c \ll 1$, the position and intensity of the fringes can be evaluated analytically. This condition applies to Newton's hair measurements, and in this case the light intensity I is given approximately by[9]

$$I \propto \frac{1}{(L+z)^2}\left[1 - 2\sqrt{\frac{c}{\pi}}\cos\left(cy'^2 - \frac{\pi}{4}\right)\frac{\sin(cy')}{cy'}\right], \tag{1}$$

where $y' = y/(a(1+z/L))$, and y is distance measured on the screen transverse to the axis of the hair. Hence, the maxima of the fringes occur when the argument of the cosine equals $(2n + 1)\pi$, where n is a positive integer, which yields,[10]

$$y = \pm\sqrt{(\lambda z(1 + z/L)(2n + 5/4)}\,. \tag{2}$$

In this approximation, we find that the position of these fringes are independent of the width a of the hair, and are located on a hyperbolic curve in the y,

z plane. This expression is similar to the result obtained from the simple two-wave interference model of Young,[11] except for the term 5/4. Thus, Newton's early attempts at a linear or ray theory for the location of the fringes were doomed to failure as he eventually came to discover[1]. In the *Opticks*, Observation 3, Newton gave measurements for the separation between the first three fringes for $L = 12$ ft., and screen distances $z = 1/2$ ft. and $z = 9$ ft. Occasionally, Newton used a prism in front of his pinhole source to select a beam of a single color, but generally he used the full spectrum of sunlight. In the following two tables, the separation of these three fringes are compared with the theoretical values obtained from Eq. (2) (all units are in inches). Since Newton used sunlight for these measurements, we take for λ a value near the peak of the spectrum (yellow) corresponding to $\lambda = 0.0000217$ inch.

$z = 1/2$ ft., $c = 0.31$		
n	$2y$	Newton
0	0.0259	0.0256 or 0.0263
1	0.0418	0.0425
2	0.0531	0.0541 or 0.0555

$z = 9$ ft., $c = 0.0128$		
n	$2y$	Newton
0	0.143	0.140
1	0.231	0.235
2	0.293	0.300

Evidently in all of these six measurements, the agreement between Newton's observations and the predictions of Fresnel's theory, Eq. (1), is very good.

Newton obtained also results for the width b of the hair's shadow, Observations 1 and 3. Newton's criterion for the termination of the shadow is not known, but as we shall show it appears to correspond to the reasonable requirement that the light intensity at the edge of the shadow be the same as in the absence of the hair. According to Eq. (1), this occurs when the argument of the cosine is set equal to $\pi/2$, which determines the width as

$$b = \sqrt{3\lambda z(1 + z/L)}\,. \tag{3}$$

In the following table, the ratio $b^2/(z(1+z/L))$ is evaluated for Newton's data, which is expected, according to Eq. (3), to be a constant with magnitude $3\lambda = 0.0000651$ in. (yellow light).

z	b	$b^2/(z(1+z/L))$
4	1/60	0.000068
6	1/54	0.000055
24	1/28	0.000046
108	1/9	0.000065
120	1/8	0.000071

In the last column of this table, the largest deviation from the mean value occurs for the measurement of b at $z = 24$ in. However, I have found that for this case, the reported value for the width of the shadow, $b = 1/28$ in., fits rather well an early unpublished linear model for the location of diffraction fringes developed by Newton[1], while the Fresnel theory predicts $b \approx 1/23$. Perhaps this is an example where Newton may have "fudged" his data in order to fit a model which he discarded later.

In a second diffraction experiment, Observation 8, Newton constructed a slit in the shape of a wedge by tilting the edges of two knives at a small angle (see A. Shapiro; Chapter 4). In his words:

> "I caused the edges of two Knives to be ground truly strait, and pricking their points into a Board so that their edges might look towards one another, and meeting near their points contain a rectilinear Angle, I fasten'd their Handles together with Pitch to make this Angle invariable. The distance of the edges of the Knives from one another at the distance of four Inches from the angular Point, where the edges of the Knives met, was the eight part of an Inch; and therefore the Angle contained by the edges was about one Degree 54 ..."

Newton assumed that any small portion of this wedge would act like a slit with parallel edges of corresponding width, enabling him in this ingenious manner to consider a full range of widths with a single experimental arrangement. It turns out that this is valid according to the wave theory of light provided that the angle θ between the knives is sufficiently small, as is the case in Newton's arrangement. In this case, according to the Fresnel theory, the light intensity on a screen parallel to the knives' wedge at a distance z

of the diffracting object is completely determined by two dimensionless parameters, $c = \pi(x\theta)^2/(\lambda z(1+z/L))$ and $\bar{y} = y(1+z/L)/(x\theta)$, where x measures distance parallel to the axis of the wedge, and y the distance perpendicular to this axis. This result implies that the wedge diffraction pattern on the screen is *invariant* with the distance z of the screen from the knives, and that it scales with $\sqrt{\lambda z(1+z/L)}$. Newton's remarkable drawing of this pattern in the *Opticks*[12] does not specify the value of z, but he would have expected the independence on z of the fringe pattern on the basis of his corpuscular or ray model of light. However, the scaling property[13] expected from the wave theory cannot be explained by this model as Newton found out from his observations.

For a slit of constant width a, it is well-known that when

$$c = \pi a^2 \left(1 + \frac{z}{L}\right) \Big/ (\lambda z) \ll 1\,,$$

the minima of the fringes occurs at

$$y = \pm n \frac{z\lambda}{a}\,, \tag{4}$$

where y is the distance transverse to the axis of the slit, and n is a positive integer. This is the familiar result obtained in elementary textbooks on optics known as Fraunhofer diffraction. In Newton's experiment with two knives forming a wedge, a is a variable given by $a = \xi\theta$, where ξ is the distance from the vertex of the wedge, and θ is the angle of aperture of the wedge. Provided that this angle is sufficiently small such that there is a range where $x \gg \sqrt{\lambda z(1+z/L)}$ while $c = \pi(x\theta)^2/\lambda z(1+z/L) \ll 1$, it can be shown that the location of the minima of the fringes on the screen, in this range of x, is given by[9]

$$y = \pm n \frac{\lambda z(1+z/L)}{x\theta}\,, \tag{5}$$

corresponding to the substitution $a = x\theta/(1+z/L)$ in Eq. (4). Hence, in this case the fringes trace a hyperbolic curve on the $x-y$ plane of the screen, a property which Newton mentions in Observation 10 in the *Opticks*,[14]

> "... and thereby know that these curve lines are Hyperbolas differing little from the conical Hyperbola."

Since in this case $y/(x\theta) = \pi/c \gg 1$, this portion of the fringes is located in the geometrical shadow of the wedge. In Observation 10, Newton also gave numerically the separations of the light minima between the first three fringes at a distance "half an Inch" from the projection on the paper of the wedge vertex. His results, 0.35, 0.65 and 0.98 inch are approximately in the Proportions 1, 2 and 3 expected from Eq. (5), but according to this equation the magnitude is off by almost a factor of two. It is implausible that Newton would have made an error of such magnitude for these relatively easy measurements, and it appears more likely that he gave incorrectly the figure 1/2 inch for the distance at which the fringe measurements where made. Elsewhere, Newton referred his measurements on the screen to values projected on the plane of the knives. These are related by the geometrical factor $(1 + z/L)$ which is 1.9 in this case, possibly accounting for the discrepancy.

For $c > 1$, symmetrical fringes on opposite sides of the wedge cross on the x axis, and Newton measured the minimum at the first crossing point for various distances z carefully. According to Fresnel's theory, this crossing occurs at a value $c \approx 19.4$ corresponding to the first minimum in the light intensity along the x axis, and correspondingly

$$x\theta = \sqrt{(c/\pi)\lambda z(1 + z/L)}\,. \tag{6}$$

Note that this crossing occurs on a hyperbolic curve in the $x - z$ plane as required by scaling for a specific pattern of the fringes on the screen.[15] In Observation 9, Newton describes his procedure,

> "When the Rays fell very obliquely upon the Ruler at the distance of the third Part of an Inch from the Knives, the dark Line between the first and second Fringe of the Shadow of the other knife met with one another, at the distance of the fifth Part of an Inch from the end of the Light which passed between the Knives at the concourse of their edges. And therefore the distance of the edges of the Knives at the meeting of these dark lines was the 160th Part of an Inch."[16]

Substituting $z = 1/3$ in Eq. (6) and neglecting $z/L \ll 1$ yields $x\theta = 0.0067$, which is in reasonable agreement with Newton's observational result $x\theta = (1/5)(1/32) = 1/160 = 0.00625$. In the next table, we reproduce in the last column, Newton's results[17] for the "distance of the edges" a corresponding to the crossing of the boundary between the first two fringes, for several values of z at $L = 8.5$ ft. Evidently, he determined a by measuring the crossing point x, which he did not report, and then applied the geometrical relation $a = x\theta/(1 + z/L)$. For comparison, the corresponding theoretical values of x and a obtained from the Fresnel theory, Eq. (6), are shown in columns 2 and 3.

z	x (Fresnel)	a (Fresnel)	a (Newton)
1.5	0.46	0.014	0.012
3.33	0.69	0.021	0.020
8.6	1.04	0.033	0.034
32	2.40	0.057	0.057
96	5.06	0.081	0.081
131	6.42	0.087	0.087

The agreement is quite impressive showing that Newton's measurements had been performed very carefully. While historians of science have commented on the accuracy of Newton's measurements before, Ref. 2, such quantitative details have not been provided previously. Alan Shapiro has shown that these measurements forced Newton to abandon his earlier attempt at a linear model for the propagation of fringes.[1] Newton concluded

> "And hence I gather that the Light which makes the Fringes upon the Paper is not the same Light at all distances of the Paper from the Knives, but when the Paper is held near the Knives, the Fringes are made by Light which passes by the edges of the Knives at less distance, and is more bent than when the Paper is held at a greater distance from the Knives."[17]

Newton never offered a model which could account quantitatively for these observations. The claim often made by historians that Newton somehow fiddled or fudged his data to fit some theoretical pre-conceptions must be re-evaluated in the light of these diffraction experiments. As Young has observed in 1801,

> "... these experiments must have been conducted without the least partiality for the system by which they will be explained ...."[18]

although some exceptions do occur, as in the case of the measurement of the hair's shadow given earlier.

I. Bernard Cohen stated that Newton's optical theory "was in fact adequate to the phenomena it attempted to explain" and that "Newton's corpuscular theory, while clearly incorrect, was nevertheless a very ingenious creation and had been fully able to explain all of the facts about light known in Newton's days."[19] Actually for diffraction fringes, Newton's theory was inconsistent. He required that light rays bend away from a narrow diffracting obstacle like a hair or a wire while bending towards the edges of a broad obstacle like a knife. Earlier he even proposed, *Opticks*, Query 3, that

> "Are not the Rays of Light in passing by the edges and sides of Bodies, bent several times backwards and forwards with a motion like that of an Eel? And do not the three Fringes of colour'd Light above-mention'd arise from three such bendings?"

Although Newton considered the possibility that the action of the edge of a knife on a light ray might even change sign with the distance of the ray from the edge, his theory violated an obvious symmetry that occurs when such a light ray passes through the middle of a slit. In this case the combined effect of such optical "forces", equidistance from opposite edges of the slit, should cancel. Hence, his observation of light minima along the axis of a slit was an insurmountable difficulty for his model which apparently Newton did not bother to address. Finally, Newton was unable to account from a fundamental principle for the discreteness of diffraction fringes. His earlier theory of fits of easy reflection and transmission, which introduced vibrations to explain fringes in such phenomena as Newton's rings and the colors in thin glass plates,[20] depended on the existence of interfaces between media with different indices of refraction, and therefore it could not be applied to diffraction fringes. In Book III of the *Opticks*, Newton did not make any attempts to relate quantitatively his measurements to his various optical models. As Rupert Hall concluded "... much of the hoped for mathematisation of the theory of colours was to remain no more than an 'idle dream' ...."[21]

Newton's optical work illustrates his empirical approach to scientific research, which he expressed as follows:

> "For the best and safest method of philosophising seems to be, first diligently to investigate the properties of things and establish them by experiments, and then to seek hypothesis to explain them."

While he failed completely to find the correct hypothesis for diffraction phenomena — the interference of light waves — his detailed and accurate experiments paved the way for the theory of Young and Fresnel. As Young aptly observed

> "The optical observations of Newton are yet unrivalled; and excepting some casual inaccuracies they only rise in our estimation as we compare them with later attempts to improve on them."[6]

Acknowledgments

I would like to thank Alan Shapiro for many fruitful discussions and clarifications.

Notes and References

1. Alan Shapiro, *Newton's Experimental Investigation of Diffraction for the Opticks: A Preliminary Study* (Chapter 4 in this volume). Shapiro is currently editing a second volume of Newton's unpublished optical papers.
2. Roger H. Stuewer, "A critical analysis of Newton's work on diffraction," *Isis* **61** (1970): 188–205.
3. M.P. Silverman and Wayne Strange, "The Newton two-knife experiment: intricacies of wedge diffraction," *American Journal of Physics* **64** (1996): 773–787. These authors consider the incident light as a plane wave and therefore their results cannot be used for numerical comparison with Newton's data except when $z/L \ll 1$. Moreover, the authors also claim incorrectly that the Fresnel theory does not justify Newton's assumption that his knives wedge is equivalent to a slit with variable width.
4. Isaac Newton, *Opticks or A Treatise of the Reflections, Refractions, Inflections and Colours of Light*, based on the 4th Ed. London, 1730 Dover Publications Inc. (1952).
5. In an early model for the fringes (see Ref. 1), Newton "quantized" an impact parameter for the scattering of light corpuscles from the hair. The fit to his measurements implied that this impact parameter was proportional to $\sqrt{n}$, where n is an integer, in approximate agreement with Eq. (2).
6. Thomas Young, "The Bakerian lecture: on the theory of light and colours," *Philosophical Transactions* **92** (1802): 12–48.
7. Thomas Young, "The Bakerian lecture: experiments and calculations relative to physical optics," *Philosophical Transactions* **94** (1804): 1–16.
8. Jed Buchwald, *The Rise of the Wave Theory of Light*, University of Chicago Press, 1989. Huygens proposed that the propagation of a wave front is obtained as an envelope of wavelets while Fresnel took the linear superposition of their amplitudes which gives rise to interference and the occurrence of fringes.
9. Michael Nauenberg (unpublished). The analytic approximation for the intensity I, Eq. (1), is in very good agreement with a numerical integration of the Fresnel integrals up to values of $c \approx 0.3$–0.4.
10. The factor $\sin(cy')/cy'$ in Eq. (1) where $cy' = \pi a y/\lambda z$, corresponds to the familiar amplitude for Fraunhofer diffraction in a slit of width a. In the case of an obstacle, it modulates the intensity of the fringes. For $L = 12$ ft. and $z = 1/2$ ft., this factor gives an appreciable effect, e.g. the fourth fringe has about $1/10$ the relative intensity of the first fringe, which may account

for the fact that Newton reported observing only three fringes. For large values of z, e.g. $z = 9$ ft., the effect of the modulation of the first few fringes is small, but in this case the overall intensity decreases by a factor $(150/252)^2 = 0.354$.

11. Young assumed that the maxima of the fringes occur when the difference d between the path length of light coming directly from the source and from the edges of the hair corresponds to an integer multiple of the wavelength of light. Neglecting the hair's width, we obtain $d = y^2/(2z(1+z/L))$, and setting $d = n\lambda$, this criterion gives

$$y = \sqrt{2n\lambda z(1 + z/L)}\,. \tag{7}$$

In Ref. 7, Young gives only the result of his numerical calculations. These are based on a somewhat more complicated expressions which takes into account the finite width of the hair.

12. *Opticks*, p. 333, see also A. Shapiro, Chapter 4 in this volume.
13. The invariance of the wedge diffraction pattern with the distance z of the screen from the wedge slit is not obvious (it was missed in Ref. 3), and depends on the condition that the wedge angle $\theta \ll 1.0$. This invariance does not occur in the case of a slit with uniform width a, where the fringe pattern varies rapidly for values of z such that $c = \pi a^2(1+z/L)/(\lambda z) > 1.0$.
14. Reference 4, p. 335. However, Newton does not give any numerical evidence in the *Opticks* that the curves traced by the fringes are hyperbolic. Some such evidence has been found in his unpublished optical papers (Alan Shapiro, private communication).
15. Young assumed that a dark fringe in the center was due to the destructive interference of a light wave from one of the edges of the knives and a wave from the center. The difference d in path length between these two waves is given by

$$d = \frac{a^2(1 + z/L)}{8z} \tag{8}$$

Young expected that d would be a constant corresponding to $d = \lambda/2$, while the correct answer in the Fresnel theory is $d = c\lambda/(8\pi)$. For $\lambda = .000217$ and $c = 19.4$, I obtain $d = .0000167$ in excellent agreement with Newton's observations for his larger values of z. In the table below, we reproduce Newton's measurement for $L = 8.5$ ft. and the corresponding values of d also given by Young, see Ref. 7.

Newton's Measurements		
z	a	d
1.5	0.012	0.0000122
3.33	0.020	0.0000155
8.6	0.034	0.0000182
32	0.057	0.0000167
96	0.081	0.0000166
131	0.087	0.0000165

16. *Opticks*, p. 331.
17. *Opticks*, p. 332.
18. Thomas Young, Ref. 6, p. 12.
19. I. Bernard Cohen in Ref. 4, p. ix.
20. Roger H. Steuwer, "Was Newton's "wave-particle duality" consistent with Newton's observations?," *Isis* **60** (1969): 392–394. In this paper, Steuwer points out that Newton's experiments with "Newton's Rings", where he replaced the air film with a water film, implied that the velocity v_ν of the "ether" vibrations caused by light in water is less than that in air. This dependence was predicted by the wave theory of Huygens, while Newton's corpuscular theory of light requires that the velocity v_p of light particles be greater in air than in water. Steuwer then claims that Newton's theory implies the relations $v_p = kn$ and $v_\nu = k/n$, where k is a constant and n is the index of refraction. Therefore, $v_p = n^2 v_\nu$ which contradicts Newton's theoretical requirement that v_p is less than v_ν, because for materials n is "always greater than unity." The flaw in Steuwer's argument is his assumption that the same constant k applies for the relation between velocity and index of refraction in the case of light particles and the corresponding ether vibrations. Assuming different constants k_p and k_ν, one finds that $v_p = (k_p/k_\nu)n^2 v_\nu$. Evidently, Newton's wave-particle theory implies k_ν/k_p is less than n^2, and not equal to unity as Steuwer assumed.
21. A. Rupert Hall, *All Was Light, An Introduction to Newton's Opticks* (Clarendon Press, Oxford, 1993).

J. Bruce Brackenridge

Newton's Dynamics: The Diagram as a Diagnostic Device

J. BRUCE BRACKENRIDGE

Department of Physics, Lawrence University
Appleton WI 54712, U.S.A.

I. Introduction

Newton developed three distinct but interrelated methods of solving direct problems of orbital motion; that is, methods for finding the nature of the force required to maintain a given orbit about a specified center of force. These three methods, which I call the polygonal, parabolic, and curvature methods, have their roots in Newton's very early work on dynamics before 1669 and all appear in his mature work after 1679. The trail is not always clearly marked, however, between the first analysis of 1664 and the publication of the first edition of the *Principia* in 1687. Moreover, it is often difficult to identify these individual methods in the detailed and intricate structure of the 1687 *Principia*, and even more difficult to do so in the expanded and revised editions of 1713 and 1726.[1]

In the first method, the polygonal method, the continuous orbit is represented by a polygon whose sides then become vanishingly small. This method was employed by Newton in his first (1664) analysis of circular motion and later in Proposition 1 of the 1687 *Principia*. In the second method, the parabolic method, an element of the continuous orbit is represented by a vanishingly small parabolic arc and can also be found in Newton's early analysis of circular motion and later in Proposition 6 of the 1687 *Principia.*[2] In the third method, the curvature method, an element of the continuous orbit is represented by a vanishing small circular arc of the circle of curvature. This method is first published as an alternate method in the revised 1713 *Principia*, but Newton did speak of it as a way of analyzing elliptical motion as early as 1664, although he did not provide a detailed description. Central to Newton's dynamics in general, and to this third method in particular, is curvature, a measure of the rate of bending of curves that Newton developed before he began his analysis of orbital motion.[3]

It is to this third method that I direct your attention. The first two methods are often given priority in discussions of Newton's dynamics, perhaps because they appear as the primary method of solving direct problems in the *Principia* (and the only method in the 1687 edition). The polygonal method is employed as Proposition 1 in providing the generalization of Kepler's law of equal areas in equal time. This result, in turn, provides a critical element in Newton's derivation in Proposition 6 of the parabolic measure of force for solving direct problems. It is curvature, however, that is basic to Newton's early and continued thoughts on dynamics. Newton states unequivocally that he did not have the generalization of Kepler's area law until after 1679, and thus he did not have the basis for applying the parabolic measure to non-circular orbits before that time. Yet he does state in 1664 that he can apply curvature to elliptical orbits, and the critical role of curvature is revealed in his diagrams and his commentary on orbital analysis.[4] I select two basic diagrams as the focus of my discussion: the diagram for Lemma 11 of Book I of Newton's *Principia* and the diagram contained in Newton's letter of 13 December 1679 to Robert Hooke. Analysis of the first diagram demonstrates the essential role of curvature in the 1687 *Principia*, even though the curvature measure does not appear explicitly as an alternate measure of force until the revised editions. Analysis of the second diagram, based upon the recent work of Michael Nauenberg, provides insights into Newton's use of curvature in the analysis of non-circular orbits before his generalization of Kepler's area law after 1679.[5]

II. Lemma 11: The Curvature Lemma and its Relationship to Proposition 6

As early as 1664, Newton had developed the concept of curvature to measure the rate of bending of curves, and by 1671 it appeared as an important element in his *Method of Series and Fluxions.* A circle, for example, has a constant rate of bending and thus the curvature is constant, while for an ellipse the curvature changes but in a uniform fashion. For the measure of curvature of a general curve at a point, Newton used the radius of the circle that has the same rate of bending at that point; that is, the circle of curvature. There is no explicit mention of curvature in the text of Lemma 11 in the 1687 edition save for a reference to "the nature of circles passing through the points A, B, G; A, b, g," although curvature is implicit in the lemma. Figure 1 is an enhanced diagram for Lemma 11 to which I have added the two circles, ABG and Abg, that are explicit in Newton's text and the circle of curvature AJ that is implicit in the text. Each circle is tangent to the general curve AbB at point A; circle ABG cuts the general curve at point B and circle Abg at point b; and they form

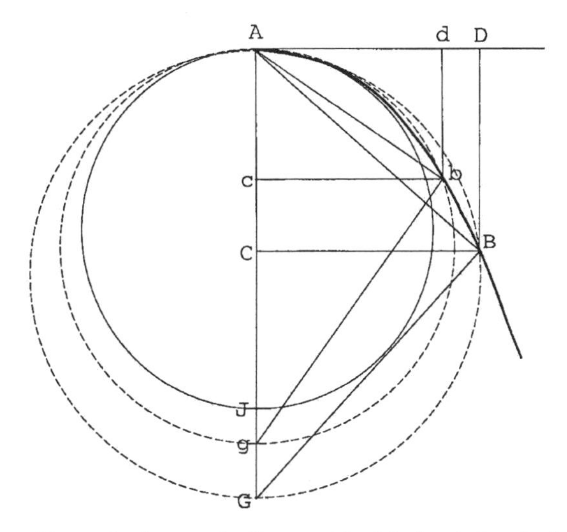

Fig. 1 Based on Newton's diagram for Lemma 11 with the circles displayed. The general curve is AbB and the circle of curvature is AJ.

their diameters at points G, g and F, respectively. Newton demonstrated that the ratio of the squares of the chords is equal to the ratio of the product of the subtense and the diameter, that is, $AB^2/Ab^2 = (BD \times AG)/(bd \times Ag)$. In the limit as B and b approach A, their diameters AG and Ag approach AJ, the diameter of curvature at point A, and the ratio (AG/Ag) approaches unity. Thus, the ratio of the squares of the chords is ultimately proportional to the ratio of the subtenses (where later the subtense is identified as being proportional to the force). Therefore, curvature is central to the demonstration of Lemma 11, and hence by extension to the propositions that call upon it.

In the first edition, in contrast to the revised editions, the only reference to Lemma 11 in the first three sections of Book I is to Proposition 4 and to Proposition 9. The former demonstrates that the force necessary to maintain uniform circular motion about a center of force at the center of the circle is directly as the square of the velocity and inversely as the radius of the circle. The latter demonstrates that the force necessary to maintain spiral motion about a center of force at the pole of the spiral is inversely as the cube of the distance.[6] A comparison from the first edition of the diagrams for Proposition 4 and Lemma 11 demonstrated a striking similarity. Moreover, Newton states in the proof of Proposition 4 in the first edition that "by Lemma 11, the nascent segment tk is to the nascent segment dc as bt^2 to bd^2,"[7] which is exactly the same language he employed in Lemma 11, that is, "the nascent segment Bd is to the nascent segment bd as AB^2 to Ab^2." The application of Lemma 11 to Proposition 4 is therefore clear and direct, both visually and textually.

In the revised edition, however, Newton simplified the demonstration of Proposition 4 so that it no longer contains a diagram. The new demonstration of Proposition 4 makes no reference to Lemma 11, instead Newton called implicitly upon relationships from Euclid.[8] There was no longer any reason, therefore, for him to retain the original form of the diagram of Lemma 11, and he was free to change it to conform to his revised Proposition 6, which now does call explicitly upon Lemma 11. But Newton did not elect to do so. In the revised edition, both the diagram and the statement of the lemma remain unchanged, except for the addition of the qualification that at the point of contact the curve must have "a finite curvature." One must look carefully, therefore, to see curvature in the background because, important as it is to Newton's thought, he did not often make explicit mention of it.[9]

In all three editions of the *Principia* (1687, 1713 and 1726), the basic algorithm for solving direct problems is given in Proposition 6. Only the third (1726) edition of the *Principia*, however, has been fully translated from the Latin, and it is that edition that is readily available to readers in English.[10] In that third edition (as in the second edition), the text of Proposition 6 calls upon Lemma 11. In the first edition, however, there is no reference in Proposition 6 to Lemma 11. Figure 2 is based upon the diagram for Proposition 6 as it appears in the first edition, and it serves to introduce Newton's parabolic method as a technique for the solution of direct problems. If there was no force acting on the body at point P, then it would continue in a straight line

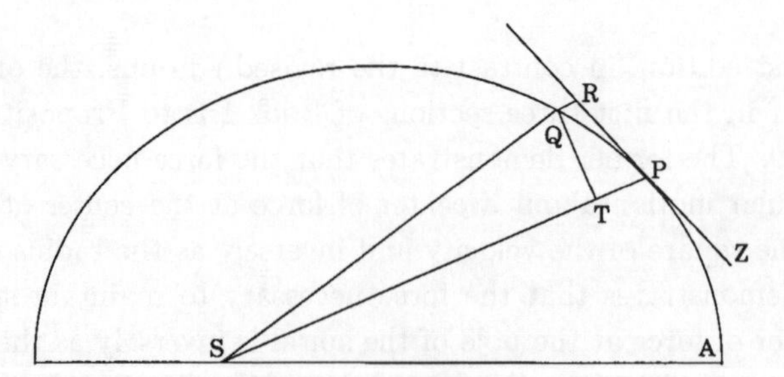

Fig. 2 Based on Newton's diagram for Proposition 6 as it appears in the first (1687) edition of the *Principia*. The body P moves on the general curve APQ with the center force at S.

along the tangent to point R. As a force directed towards S does act upon the body, it moves along the curve to point Q. In the limit as the point Q approaches the point P, Newton argued that the force is proportional to the displacement QR and inversely proportional to the square of the time. In Proposition 1, Newton demonstrated that the time is proportional to the area SP $\times$ QT, thus the reciprocal of the force is proportional to $\mathrm{SP}^2 \times \mathrm{QT}^2/\mathrm{QR}$; that is, the parabolic measure.

In the first edition the dependence of the force upon the square of the time is referred to Lemma 10, which states precisely what Proposition 6 requires; that is, "the nascent line segment QR is... given the force, as the square of the time (by Lemma 10)."[11] In the revised edition, however, the primary reference in Proposition 6 to the dependency of the force upon the square of the time is to Corollaries 2 and 3 of Lemma 11 and only secondarily to Lemma 10; that is, "the same thing may also be easily demonstrated by Corollary 4 of Lemma 10."[12] Figure 3 is based upon Newton's revised diagram for Proposition 6, which I have enhanced by the addition of the circle of curvature PV.[13] Comparison with the original diagram (in Fig. 2) will show that the most obvious change is the addition of the dotted line YS, which passes through the force center S and is normal to the tangent YPZ. A more subtle but even more significant change in the figure is the extension of the line of force SP through

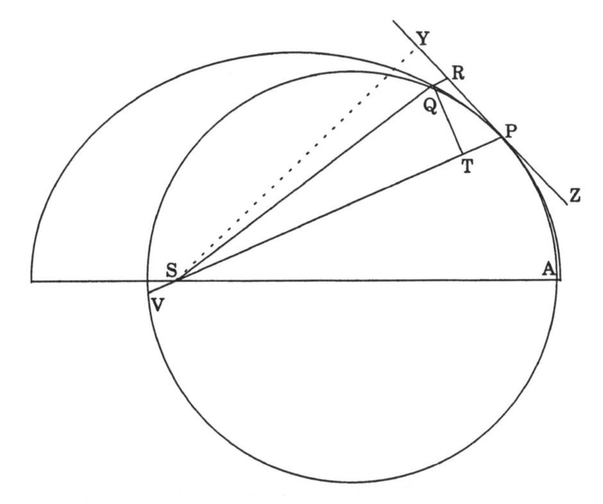

Fig. 3 Based on Newton's revised diagram for Proposition 6 as it appears in the 1713 and 1726 editions of the *Principia* but with the circle of curvature PQV added.

the force center S to a point V, where the line PV is identified as the chord of curvature from P through the center of force S. In this revised Proposition 6, Newton still derived the parabolic measure of force, $QR/(SP^2 \times QT^2)$, but in addition he derived an alternate measure of force, $1/(SY^2 \times PV)$, which is clearly dependent upon curvature because PV is the chord of curvature through the point S, the center of force.

The relationship of the latter (the curvature measure) to the former (the parabolic measure) can be seen by applying the curvature technique of Lemma 11 to the revised diagram of Proposition 6. Figure 4 is an enhanced version of Newton's revised diagram for Proposition 6 with the addition of the circle of curvature PV and an auxiliary circle PQU tangent at *P* and passing through *Q*. Following the argument from Lemma 11, as the point Q approaches the point P, then the auxiliary circle PQU approaches the circle of curvature PV (as in Lemma 11, when the point B approaches the point A, then the auxiliary circles ABG and Abg approach the circle of curvature AJ). Thus, one can employ Euclidean relationships that are valid for the auxiliary circle to obtain exact relationships for the general curve; that is, for the circle of curvature. In particular, Proposition 36 from Book III of Euclid's *Elements* is directly applicable to Newton's revised diagram. This Euclidean relationship

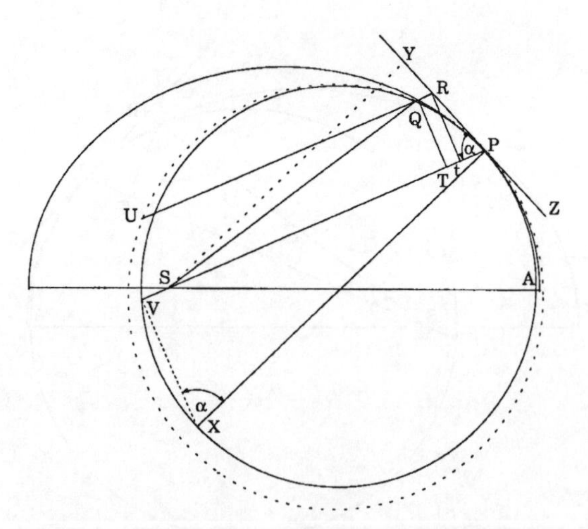

Fig. 4 An enhanced version of Newton's revised diagram for Proposition 6. PVX is the circle of curvature and PQU is an auxiliary circle. As the point Q approaches the point P. the the auxiliary chord QU approaches the chord of curvature PV.

is one that Newton employs elsewhere in the *Principia* (for example, Propositions 4 and 7 of Book I of the 1687 *Principia*), and often without any explicit reference. When applied to the auxiliary circle PQU in Fig. 4, Euclid (III, 36) demonstrates that RU:RP::RP:QR.[14]

The challenge of solving direct problems using the paradigms of Proposition 6 is to express the dependance of the force in terms of spatial parameters (such as the radius SP) and other orbital parameters (such as the diameter of a circular orbit). The key to such solutions is the evaluation of the limiting value of the ratio QT^2/QR from the parabolic measure of force $QR/(SP^2 \times QT^2)$ as the advanced orbital point Q approaches the initial orbital point P; that is, $\text{Lim}_{Q->P}(QT^2/QR)$. In the first edition of the *Principia*, Newton did not supply much insight into the nature of this ratio, but in the second (and third) edition, the revision of the demonstration of Proposition 6 permits one to understand the limiting value of this important ratio. From Euclid (III, 36) one has for the auxiliary circle the following relationship:

$$\text{RU} : \text{RP} :: \text{RP} : \text{QR} \text{ or the equivalent, } \text{RU} = \text{RP}^2/\text{QR}$$

As the lines RQU and PTV are parallel, then $Rt = QT$ and thus $RP = QT/\sin(\alpha)$, where α is the angle RPT. Thus, in the limit as Q approaches P, one obtains from curvature and Euclid (III, 36) the following exact relationship:

$$\lim_{Q->P}(\text{RU}) = (1/\sin^2\alpha) \lim_{Q->P}(\text{QT}^2/\text{QR}) = \text{PV}$$

Because the triangles RPt and PXV are similar, then the chord of curvature PV equals the diameter of curvature $PX(= 2\rho)$ multiplied by $\sin\alpha$, where ρ is the radius of curvature at point P, or

$$\lim_{Q->P}(\text{QT}^2/\text{QR}) = 2\rho\sin^3\alpha$$

Thus, even though both QT and QR vanish in the limit as Q approaches P, the ratio QT^2/QR remains finite, being dependent upon the radius of curvature ρ and the angle α for a given point P. From Proposition 6, the force is inversely proportional to SP^2 multiplied by the limit of the ratio QT^2/QR, thus

$$(\text{force})^{-1} \propto \text{SP}^2 \times \lim_{Q->P}(\text{QT}^2/\text{QR}) = 2\rho\text{SP}^2\sin^3\alpha$$

Since $PV = 2\rho\sin\alpha$ and $SY^2 = SP^2\sin^2\alpha$, then the measure of force is also given by $(\text{force})^{-1} \propto PV \times SY^2$, which is the alternate measure of force (the curvature measure) that Newton adds to Proposition 6 in the revised *Principia*.

The application of the measure of force to the four direct problems that Newton addresses in Book I of the 1687 *Principia* is particularly clear in the form shown above, that is, $F^{-1} \propto 2\rho r^2 \sin^3 \alpha$ where r is the radius vector. The following table contains the solutions for those direct problems. In Proposition 7, the object P orbits on a circle of diameter 2R and the center of force S is located on the circumference of the circle, i.e. $r = \mathrm{SP}$. Newton demonstrated that the force is inversely proportional to the fifth power of the radius SP. In Proposition 9, the object P orbits on an equal angular spiral with the center of force S located at the pole of the spiral. A property of the spiral is that the sine of the constant angle α between the polar radius SP and the tangent to the curve is equal to SP/ρ. Newton demonstrated that the force is inversely proportional to the inverse cube of the radius SP. The results are summarized in Table 1.

Table 1

	$\sin\alpha$	ρ = radius of curvature	$2\rho\sin^3\alpha$	$F^{-1} \propto 2\rho r^2 \sin^3\alpha$
Proposition 7	SP/2R	Constant = R	$\mathrm{SP}^3/(8\mathrm{R}^2)$	$\mathrm{F}^{-1} \propto \mathrm{SP}^5/(8\mathrm{R}^2)$
Proposition 9	Constant = k	$\mathrm{SP}/\sin\alpha = \mathrm{SP}/k$	$\mathrm{SP}(2k^2)$	$\mathrm{F}^{-1} \propto \mathrm{SP}^3(2k^2)$

Figure 5 is based upon the diagram from the 1687 *Principia* for Propositions 10 and 11. I have removed a number of lines and points, however, and have added the circle of curvature PXV and the angles α_{C}, α_{S}, and β.[15] In these propositions the object P orbits on an ellipse, with the center of force at the center C of the ellipse in Proposition 10 and at the focus S of the ellipse in Proposition 11. In the revised Proposition 10, Newton demonstrated that the chord of curvature PV through the center C of the ellipse is equal to the conjugate *latus rectum* $\mathrm{L_c}$ of the conjugate diameters DCK and PCG (where the conjugate diameter DCK is constructed parallel to the tangent ZPR and $\mathrm{L_c}$ is defined by the ratio $\mathrm{L_c}$:2CD::2CD:2PC). As angles PVX, YPX and PFC are right angles, then $\alpha_{\mathrm{C}} = 90° - \beta$ = angle PCF = angle PXV. Thus, $\mathrm{PV} = \mathrm{PX}\sin\alpha_{\mathrm{C}} = 2\rho\sin\alpha_{\mathrm{C}} = \mathrm{L_C} = 2\mathrm{CD}^2/\mathrm{PC}$, or $\rho = (\mathrm{CD}^2/\mathrm{PC})/\sin\alpha_{\mathrm{C}}$. For Proposition 10, $\sin\alpha_{\mathrm{C}} = \mathrm{PF}/\mathrm{PC}$, and thus the radius of curvature at point P is $\rho = \mathrm{CD}^2/\mathrm{PF}$. In Lemma 12, Newton established that the area $\mathrm{CD} \times \mathrm{PF}$ is a constant of the ellipse, thus the force is directly as the radius $r = \mathrm{PC}$. For Proposition 11, $\sin\alpha_{\mathrm{S}} = \mathrm{PF}/\mathrm{PE}$ and Newton demonstrated that the line $\mathrm{PE} = \mathrm{AC}$, where AC is the given semi-major axis of the ellipse. Also, the area

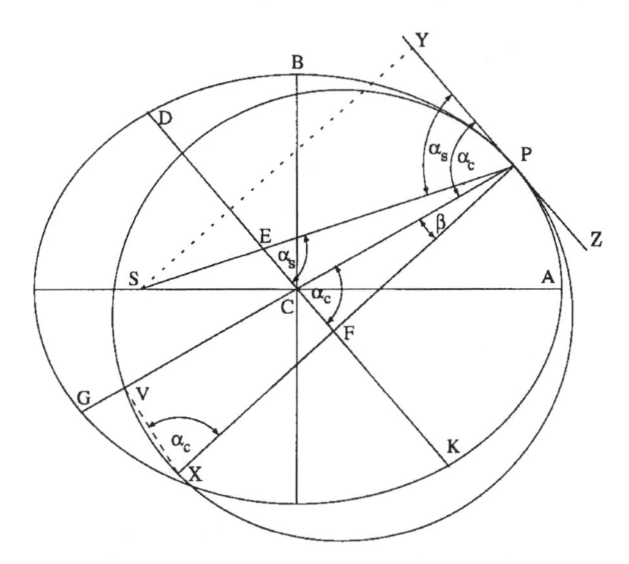

Fig. 5 Based on Newton's 1687 diagram for Propositions 10 and 11. The body P moves on the ellipse APB with the center of force at C for Proposition 10 and at the focus S for Proposition 11. PVX is the circle of curvature at the point P.

Table 2

	$\sin\alpha$	$\rho =$ radius of curvature	$2\rho\sin^3\alpha$	$F^{-1} \propto 2\rho r^2 \sin^3\alpha$
Proposition 10	PE/PC	CD^2/PF	$2(CD \times PF)^2/PC^3$	$F^{-1} \propto PC^{-1}2area^2$
Proposition 11	PF/PE	CD^2/PF	$2(CD \times PE^2/PE^3$	$F^{-1} \propto SP^2(L_p)$

$CD \times PF =$ area $AC \times BC$. Thus, $2(CD \times PF)^2/PE^3 = 2(AC \times BC)^2/AC^3 = L_p$, where L_p is the given principal *latus rectum* $2BC^2/AC$, and therefore the force is inversely as the square of the radius $r = SP$. The results are summarized in Table 2.

Thus, if one replaces an element of arc of a general curve with an element of the circle of curvature at that point, then Proposition 36 of Book III of Euclid's *Elements* can be employed to derive both the results of Lemma 11 and the two measures of force from Proposition 6. Moreover, it can be employed to obtain the solutions to the direct problems of Book I of the 1687 *Principia*. This demonstration indicates the relationship between Lemma 11 and Proposition 6, and it makes explicit the role of curvature in Newton's solution of the direct problems.

III. The Diagram from Newton's 13 December 1679 Letter to Hooke and its Relationship to Curvature

This diagram of 1679 is one that has been the object of considerable scholarly concern for more than a century. It is part of a trilogy of diagrams that are found in the correspondence of Isaac Newton and Robert Hooke at the close of 1679; specifically, 28 November 1679 Newton to Hooke, 9 December 1679 Hooke to Newton, and finally 13 December 1679 Newton to Hooke.[16] What is of interest about Newton's 13 December 1679 diagram is the information that it contains that is not available in the text that surrounds it. The problems raised by both text and diagram are many, as are the number of scholars who have struggled with them.

An early reference to this letter appeared in W.W. Rouse Ball's essay of 1893 on Newton's diagram. Rouse Ball reprinted the full text of Newton's initial response to Hooke from Newton's original holograph letter in the Cambridge University Library. He reported, however, that the two other letters were lost. He gave that portion of Hooke's "lost letter of 9 December" that was read into the minutes of the meeting of the Royal Society, and Hooke's reply to Newton's "lost letter of 13 December," from which he drew some inferences concerning its contents.[17] He never saw the diagram, however. Fortunately, both of these letters with diagrams were ultimately found. Newton's "lost letter" appeared in a sale at Sotheby's in 1904 and was purchased by the British Library. Hooke's "lost letter" appeared in a sale at Sotheby's in 1918, and was finally acquired by the Yale University Library.[18]

Figure 6 is the page of the 13 December letter containing the diagram. From the text, it appears that an object orbits inside the earth subject to a gravitational force that remains *constant*, even though the distance from the center of the earth changes. Moreover, the text speaks of a *centrifugal* force that "alternately overbalances" the constant gravity. Finally, Newton referred to his method of analysis as "the method of indivisibles," but he gave no explanation beyond tantalizing Hooke with the statement that he, Newton, "might add something about its description by points *quam proximé*," but declined to do so because it was "of no great moment." This text was first published in 1929 by Jean Pelseneer with his attempt to replicate the diagram.[19] The first photographic reproduction of the actual diagram was published in 1960 by Johs Lohne, with its size reduced by 4/5.[20] The only other photographs of the diagram appeared in publications by Curtis Wilson in 1989 and by Michael

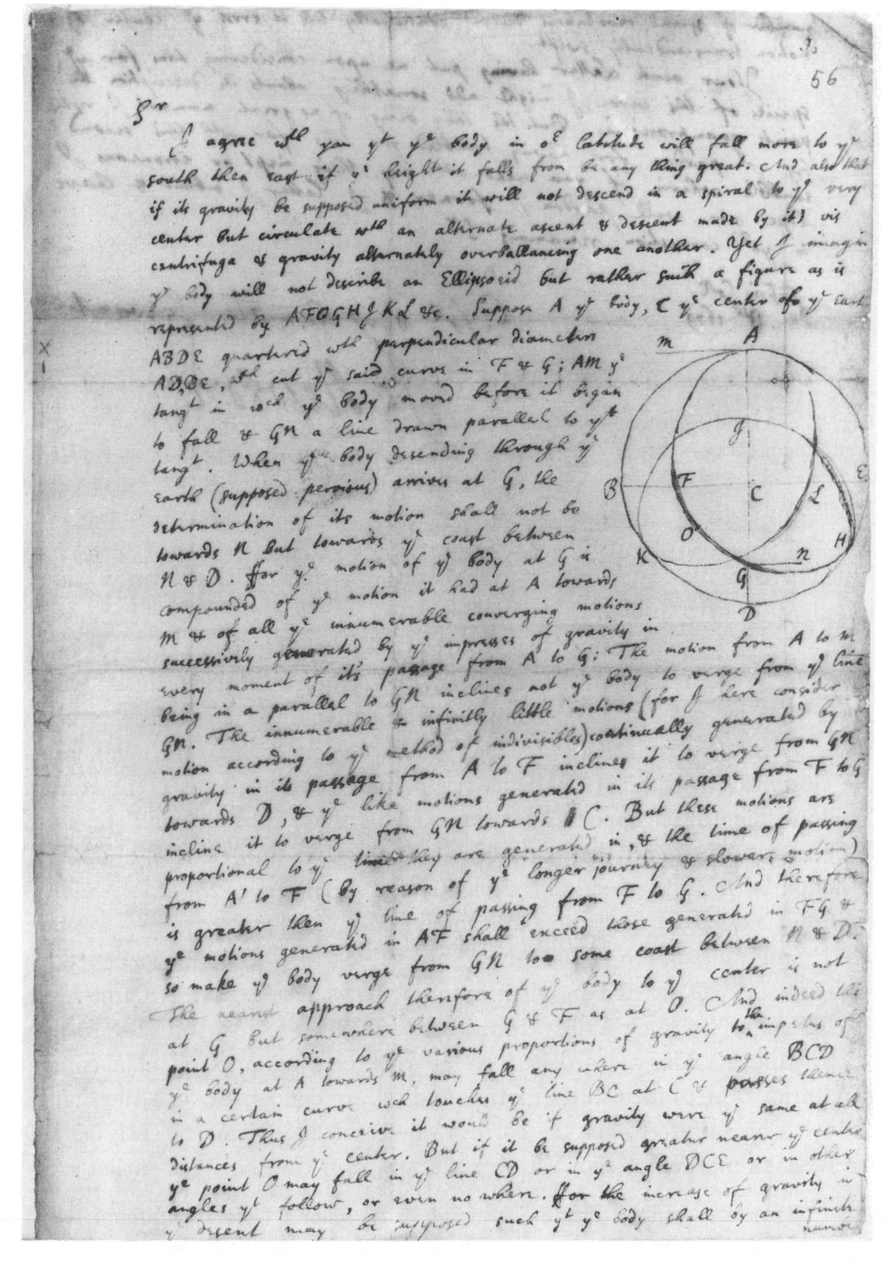

56

Sr

I agree wth you yt ye body in or latitude will fall more to ye south then east if ye height it falls from be any thing great. And also that if its gravity be supposed uniform it will not descend in a spiral to ye very center but circulate wth an alternate ascent & descent made by it's vis centrifuga & gravity alternately overballancing one another. Yet I imagin ye body will not describe an Ellipsoid but rather such a figure as is represented by AFOGHIKL &c. Suppose A ye body, C ye center of ye Earth ABDE quartered wth perpendicular diameters AD, BE, wch cut ye said curve in F & G; AM ye tangt in wch ye body moved before it began to fall & GN a line drawn parallel to yt tangt. When ye body descending through ye Earth (supposed pervious) arrives at G, the determination of its motion shall not be towards N but towards ye coast between N & D. For ye motion of ye body at G is compounded of ye motion it had at A towards M & of all ye innumerable converging motions successively generated by ye impresses of gravity in every moment of it's passage from A to G: The motion from A to M being in a parallel to GN inclines not ye body to verge from ye line GN. The innumerable & infinitly little motions (for I here consider motion according to ye method of indivisibles) continually generated by gravity in its passage from A to F incline it to verge from GN towards D, & ye like motions generated in its passage from F to G incline it to verge from GN towards C. But these motions are proportional to ye time they are generated in, & the time of passing from A to F (by reason of ye longer journey & slower motion) is greater then ye time of passing from F to G. And therefore ye motions generated in AF shall exceed those generated in FG & so make ye body verge from GN to some coast between N & D. The nearest approach therefore of ye body to ye center is not at G but somewhere between G & F as at O. And indeed the point O, according to ye various proportions of gravity to the impetus of ye body at A towards M, may fall any where in ye angle BCD in a certain curve wch touches ye line BC at C & passes thence to D. Thus I conceive it would be if gravity were ye same at all distances from ye center. But if it be supposed greater nearer ye center ye point O may fall in ye line CD or in ye angle DCE or in other angles yt follow, or even no where. For the increase of gravity in ye descent may be supposed such yt ye body shall by an infinite number

Fig. 6 The page of Newton's 13 December 1679 letter to Hooke that contained the diagram. Copyright © The British Library. Reproduced by permission.

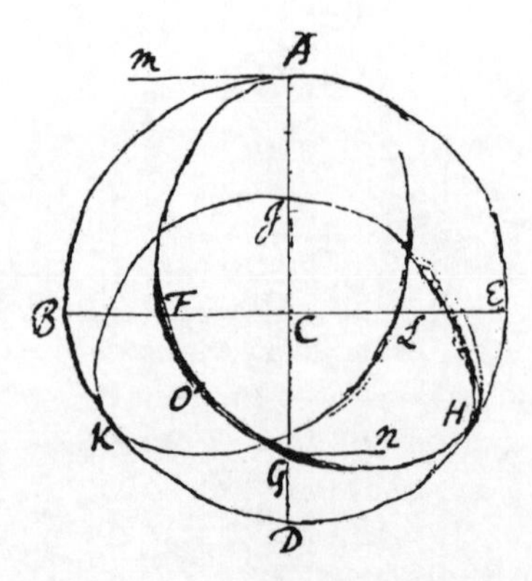

Fig. 7 An enlargement of Newton's diagram from the 13 December 1679 letter. A body orbits the center of force C. The motion begins at apogee A with speed m, moves through perigee O and continues to the next apogee H and then to the third apogee K. The curve is not closed.

Nauenberg in 1994.[21] All other commentators generated their own versions of the diagram with limited degrees of success.

Figure 7 is an enlargement of Newton's diagram. The orbit starts at apogee A, which is the maximum distance from the center C, with an initial velocity along the tangent Am, and continues through perigee at O, the minimum distance from the point C. It reaches the next apogee where the orbit touches the circumscribed circle ABDE at the point H and again at point K. From Newton's diagram it appears that the angle between apogee A and the perigee O, is about 130°, and the angle from perigee O to apogee H is about 110°, although both angles should be the same and should be considerably less than Newton's values. In fact, the angle should lie somewhere between a minimum of 90° and a maximum of 104°; Newton's value averages about 120°.

These evident textual and diagrammatic flaws have led some commentators to challenge Newton's ability to address, much less to solve, the challenge set by Hooke. Many are disturbed by Newton's assumption of a constant gravitational force for orbital motion and all are concerned with the diagram and with his failure to get the correct values for angles. As recently as 1990, the Russian mathematician V.I. Arnol'd joined in the chorus of earlier critics.

> "Newton... replied on 13 December with a long letter containing a lengthy discussion and clearly showing that at the time Newton did not know what the trajectory of the ball should look like.... This letter contains among other mistakes, an impossible picture of an orbit....

> The angle between the pericentre and the apocentre is 120° (it should belong to the interval $\pi/2[90°]$ and $\pi/\sqrt{3}[104°]$) and the orbit is clearly asymmetric."[22]

I am now convinced, however, that at last a full explanation for this diagram has been provided by Nauenberg.[23] Moreover, the final resolution of the perceived flaws of the diagram provides more than a correction for his angles or the reason for his choice of a constant force; it provides insights into Newton's dynamics during the decade previous to the letters of 1679: a period for which the written record provides scant information. In particular, it provides insight into the role that curvature played in Newton's dynamics between the early 1660s and its appearance in his revision of the *Principia*. Newton's projected relationship of curvature to dynamics first appears in his *Waste Book* in late 1664 or early 1665, while he was still a student at Trinity College, Cambridge.

> "If the body *b* moved in an Ellipsis then its force in each point (if its motion in that point be given) may be found by a tangent circle of Equal crookedness with that point of the Ellipsis."[24]

In this cryptic statement, Newton indicated that he knew how to use curvature ("crookedness") to obtain the force necessary to move a body in an ellipse; this is the solution to the so-called direct Kepler problem. The related problem addressed by Newton in his letter of 13 December 1679 is the inverse of such direct problems: that is, given the nature of the force (in this case a constant), one is required to find the path. Newton did not append a demonstration of his curvature technique to this early cryptic statement, and as a consequence, it is generally held that curvature did not play a major role in his analysis until Newton proposed the revisions of the 1690s, following the publication of the 1687 edition of the *Principia*. However, it is clear from the analysis of Lemma 11, discussed above, and from other isolated examples that curvature was an active element in the 1687 *Principia*. Moreover, the final resolution of the diagram in the letter of 13 December 1679 reveals the central and basic role curvature played in Newton's dynamics during the preceding fifteen years. That final resolution of the diagram provides a link between the cryptic curvature statement of 1665 and the alternate curvature solutions of 1690.[25]

Rouse Ball's discussion in 1893 of a portion of Newton's text was expanded in 1929 when Pelseneer published the full text of Newton's 13 December letter with a commentary and with Pelseneer's replication of the drawing of Newton's diagram, the first of a number of attempts to replicate the diagram.[26]

Included in Pelseneer's 1929 article was a modern analytical expression for the correct angle between apogee and perigee.[27] In 1960, Lohne published a photograph of Newton's original drawing along with his own version of the "correct" diagram, generated by a numerical calculation employing curvature, a calculation that he left for the most part unexplained.[28] Four years later, in 1964, Whiteside published an article that included further discussion of the diagram and in which he provided details of the modern analytical solution given by Pelseneer in 1929.[29] In 1965, John Herivel published the text with a very limited commentary and a very poor attempt to replicate the diagram,[30] and in 1971 R.S. Westfall also published the text with commentary and yet one more unsuccessful attempt at replicating the diagram.[31]

In 1974, Whiteside published the sixth volume of the eight volumes of *The Mathematical Papers of Isaac Newton.* No other work has played a more important role in our understanding of the development of dynamics than this volume, both for its presentation of Newton's published and unpublished writings and for its extensive notes, which are in themselves an introduction to late seventeenth and early eighteenth century dynamics. Of particular relevance to the 13 December diagram are the surviving portions of Newton's initial revision of the tract *On Motion*, which served as an intermediary text between the first draft of 1684 and the final draft of 1685/1686, and which was to be transcribed into the press copy of the 1687 *Principia.* This initial revision contains a scholium to the early propositions on conic motion in which Newton returned to the subject discussed in his 13 December letter to Hooke, but unfortunately Newton did not provide an accompanying diagram.[32] In his notes to this scholium, Whiteside gave an extended discussion of the modern analytical solution that appeared in his 1964 paper and a corrected diagram of the curve generated by that solution.[33] In this scholium, Newton gave an approximate value of 110° for the angle between apogee and perigee for a constant force — an angle much closer to the maximum value of 104° and much less than the angle of 130° shown in his diagram of 1679. Moreover, in this scholium, Newton gave a reasonable approximation for the angle if the force were proportional to the reciprocal of the distance. This scholium was deleted, however, from the final version of the 1687 *Principia.* The reason for that deletion may be found in the later Proposition 45 of Book I of that 1687 *Principia*, which contains Newton's calculation of an exact upper limit for a number of different forces, including the correct value of 103°55′ for the maximum value of the angle in his 13 December diagram.[34] Presumably, once Newton had developed an analytical technique for calculating the exact limit in Proposition 45, he no longer needed the results of his numerical technique that was responsible for, but not shown in, the discarded scholium to the earlier proposition.

In the two decades following the publication of Volume 6 of Newton's mathematical papers, there have been a number of commentators on the text and diagram of Newton's letter of 13 December 1679. With the exception of two commentators who used a photographic reproduction of Newton's actual diagram (to be discussed below), all others used diagrams similar to those that appear either in Koyre's article of 1952 or Turnbull's correspondence of 1960 and none were successful in replicating Newton's flawed diagram. Figure 8 is a comparison of Newton's original diagram with those of Koyre and Turnbull. The most obvious difference is with Turnbull's diagram, in which the curve is closed, but there are other important differences that become apparent upon closer inspection.

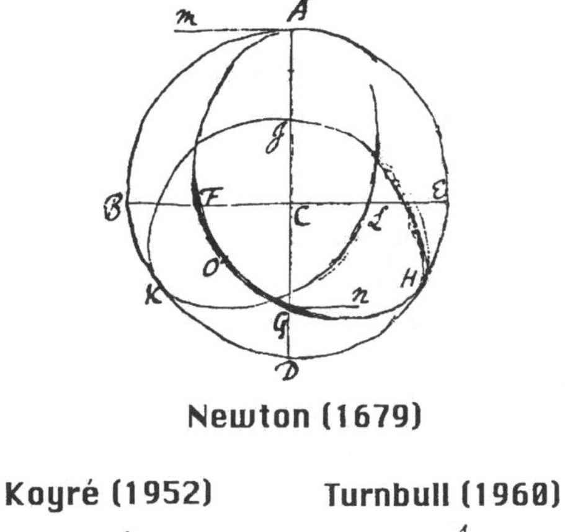

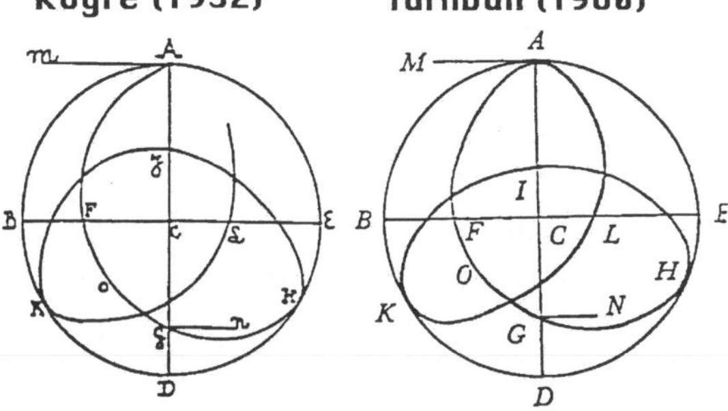

Fig. 8 A comparison of Newton's original diagram with those of Koyré and Turnbull.

In 1989, Curtis Wilson included a photographic reproduction of Newton's diagram in his discussion of the 13 December letter, but he made no attempt to use the diagram except as a general guide to Newton's analysis. He was concerned with Newton's "description of points *quam proxime*," and presented what he thought might be a likely interpretation. It consisted of applying the polygonal technique that Newton employed in deriving the areal law in Proposition 1 of the *Principia.* Wilson noted, however, that "the method we are suggesting is graphical and so but crudely approximative." He made no attempt to generate a version of Newton's diagram, and closed with the observation that Newton had obtained an angle between apogees that is "impossibly large."[35]

In 1994, Michael Nauenberg also employed a photographic reproduction of Newton's diagram, but he called upon a detailed analysis of the diagram employing curvature to suggest an interpretation of Newton's "description of points *quam proxime*" that is capable of generating the diagram. Moreover, he illuminated a drawing error that Newton made in the diagram that is responsible for the impossibly large value of the angle between apogees.[36] We next return to the comparison of diagrams in Fig. 8 and note with Nauenberg the following features:

(1) The approximately orthogonal axes AD and BE drawn on this diagram do not divide the diagram into equal quadrants, although Koyrè, Turnbull, and all other published replications of this diagram have ignored this important fact.
(2) The outer closed curve ABKDHEA is not actually a circle, although it has invariably been drawn as a precise circle in replications of it, as do Koyrè and Turnbull.
(3) Only the segment KDHE of the outer closed curve is part of a circle centered at C.
(4) The curve AFOGH is mirror symmetric.

This property of mirror symmetry can be demonstrated by taking a second transparency of Newton's diagram, reversing it, and putting it over the original diagram in Fig. 7; that is, making the second a reversed mirror image of the first. If one aligns the reversed apogee H with the original apogee A, and *vice versa*, then the curve AFOGH and the reversed curve HGOFA will be identical, thus displaying mirror symmetry.

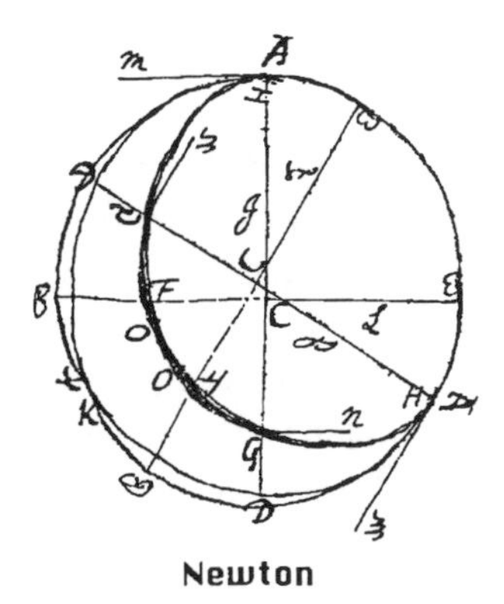

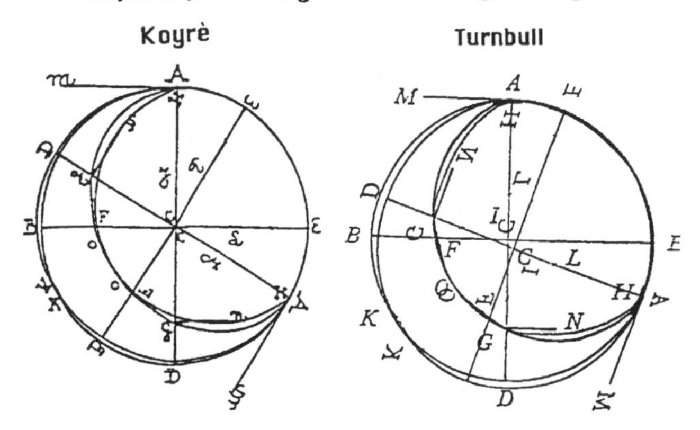

Fig. 9 A comparison of the superposition of the segment AFOGH and its mirror image for Newton's original diagram with those of Koyré and Turnbull. Newton's curve displays mirror symmetry, the others do not.

Figure 9 is a comparison of the superposition of the segment AFOGH and its mirror image for Newton's original diagram and those of Koyrè and Turnbull, where all of the curve beyond H has been removed in order to see more clearly the segment AFOGH and its mirror image.[37] Newton's diagram demonstrates this property of mirror symmetry, a feature he had known as fundamental to orbital dynamics. Inspection of the other two diagrams in the region of the letter N shows clearly that they do not have the mirror symmetry found in Newton's diagram; that is, the two curves are not identical. This property is missing, moreover, in all other attempts to replicate Newton's diagram.[38] Thus, the scholars who did not examine the original diagram with mirror symmetry, but replicated diagrams devoid of the mirror symmetry, were severely handicapped in their analysis.

Further inspection of Fig. 9 yielded yet another important feature of Newton's diagram: the center C of the reversed mirror image is shifted relative to the center C of the original diagram. This displacement of the centers suggests that Newton produced the curve by calculating a portion of it, say AO between apogee A and perigee O, and then generated the remaining portion, say OH between perigee O and apogee H, by reflection; that is, there exists a suitable rotation of a mirror reflection of a segment of the diagram that will produce the nearly perfect mirror image of the portion of the orbit AFOGH that is found in Newton's original diagram.

Figure 10 is Newton's diagram replicated by a suitable adjustment to produce the mirror symmetry.[39] The lower part of the diagram is the segment of Newton's diagram subtended by the outer closed curve BDH about the center C, and the upper segment is the mirror reflection of that lower segment rotated with its center C_S shifted up and to the left relative to the center C of the lower segment. Nauenberg concluded that Newton "made an error in shifting the center C relative to the center C_S when he produced the diagram and that he

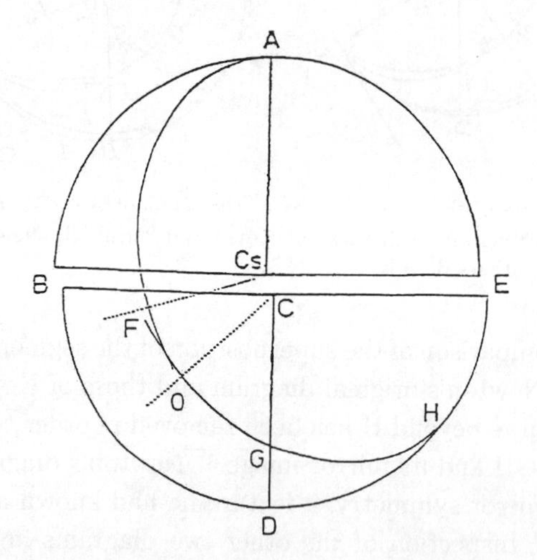

Fig. 10 Nauenberg's reconstruction for Newton's diagram, which is replicated by a suitable adjustment to produce the mirror symmetry. The lower part of the diagram is the segment of Newton's diagram subtended by the outer closed curve BDH about the center C, and the upper segment is the mirror reflection of that lower segment rotated with its center C_S shifted up and to the left relative to the center C of the lower segment.

then incorrectly adjusted the rotation in order to join these two segments of the orbital curve as smoothly as possible."[40] In support of that conjecture, he noted that Newton had patched up the juncture of the two curves in the arc FOG with multiple pen strokes (see Fig. 7), in contrast to the remainder of the curve which is smooth. (It seems very appropriate that this section is labeled FOG.)

There still remains the question of how Newton calculated the segment of the curve between apogee A or H and perigee O, and what the full correctly drawn curve AFOGH should look like. In Sec. III of his paper, Nauenberg gave an approximate geometrical (graphical) construction for a section of the curve that is based on the curvature relation implicit in Newton's 1664 "cryptic" remark, and on the relationship vdv/dr = acceleration, which when integrated leads to what is now called the principle of energy conservation.[41] Employing this curvature method, one can calculate the section of the orbit about a center C between the maximum radial distance at points A or H and the minimum radial distance at point O; that is, the portions AFO or OGH of the orbit.

Figure 11 is a diagram about a single center C: the arc AFO is obtained by the numerical curvature method and the arc OGH is obtained by reflective symmetry about the axis OC. One begins, for example, at the apogean point A with a constant acceleration a_0 directed toward the center C (for simplicity set $a_0 = 1$, and $r_0 = 1$) and a tangential velocity $v_0 < 1$ and thus calculates the radius of curvature ρ_0 at this point ($\rho_0 = v_0^2/a_0 \sin\theta_0$, where θ_0 is the angle between the tangent and the radius directed to the force center C, here 90° for the initial step, or $\rho_0 < 1$). The first segment of the curve is an arc of this circle of curvature (say for an arbitrary angle $\theta = 22°$, which divides Newton's approximate angle of 220° between apogee and perigee into ten steps), and then repeat the procedure by obtaining a new velocity and a new angle and thus a new radius of curvature for each 22° arc.[42] In contrast to Newton's diagram, both segments of this diagram are about the same center C. The angle between apogee A and perigee O is about 110°, a reasonable approximation to the maximum value of 104° given by Newton in Proposition 45, Book I. The outer curve is a perfect circle and the four quadrants are equal in size. As the size of the angular steps is reduced, the diagram approaches more closely the exact curve.

Consider once again the concerns that were raised by some critics of this diagram. Many scholars are disturbed by Newton's assumption of a constant gravitational force for orbital motion and all are concerned with the asymmetric arc

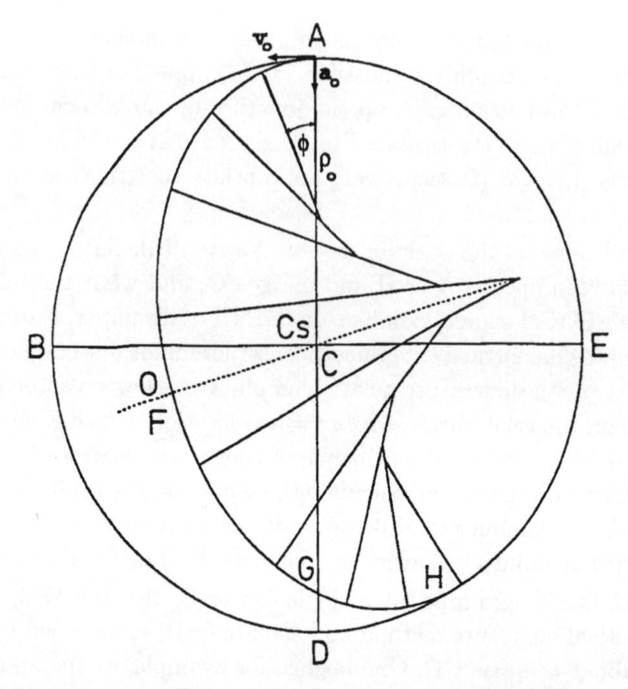

Fig. 11 Nauenberg's reconstruction of the diagram with a single center C(= C_S): the arc AFO is obtained by the numerical curvature method and the arc OGH is obtained by reflective symmetry about the axis OC.

AOH in the diagram and with his failure to get the correct values for angles. Moreover, there was no indication of what Newton's "method of indivisibles" was, or how the approximation implied by "points *quam proximé*" [points "as closely as possible"] applied to it. Given this reconstruction, however, these questions are answered. The method consists of approximating elements of the curve by elements of the circle of curvature, a technique that will be employed in the analytical solutions of the revised *Principia.* The choice of a constant force for the orbiting body is just an initial choice of the simplest example for a numerical solution; it is later to be replaced by forces dependent upon the distance or by the inverse square of the distance or by whatever complexity is desired. Finally, the angles between the apses are correct within the limits of the numerical technique.

The analysis of the diagram also sheds light on the question of what Hooke learnt from Newton as a result of this correspondence. One school of thought, leaning on the contemporary distinction between centrifugal and centripetal force arising from an analysis of rotating coordinate systems, claims that Hooke taught Newton to understand that distinction.

> "Under the tutelage of Hooke, he [Newton] bid centrifugal force *adieu* and turned down the path that would lead him eventually to the *Principia*."[43]

But that view ignores that Newton continued to use centrifugal force in his discussion of motion even after the publication of the *Principia*.[44] From the perspective of the insights gained from the understanding of the 13 December 1679 diagram, Hooke obtains credit for something quite different! In his first recorded analysis of uniform circular motion, Newton employed a series of discrete impulses to generate a polygon that approximated the circular orbit; eventually requiring that the time between impulses becomes infinitely small: the polygonal measure. In this measure the force is directed toward the center of force that is at the center of the circle, and thus the force is normal to the tangent of the circle. In a second early analysis of uniform circular motion, Newton replaced the discrete impulse with a constant continuous force for a vanishingly small time — the parabolic measure. Here, the force is also directed normal to the tangent of the circle. Finally, in the curvature method that Newton used in the 13 December 1679 analysis discussed above, attention is also given to the continuous component of force normal to the tangent that is directed toward the center of curvature, *but that component is no longer directed toward the center of force.*

Hooke's suggestion of 1679 that planetary motions are "compounded of a direct motion by the tangent and an attractive motion towards the central body" led Newton back to the polygonal method of impulses, previously used in circular motion, for analysis of non-circular motion. Moreover, Hooke directed attention to the radial change of motion due to the impulsive *radial* force directed toward the force center; that is, the *centripetal* or center-seeking force, rather than that due to the normal component force.[45] The analysis that Hooke suggested is precisely the one Newton employed to obtain Kepler's law of equal areas in equal times, which occupies the place of honor as the first proposition in both *De Motu* and the *Principia*. It is that proposition that permitted Newton to generate the measure of force in Proposition 6 in the 1687 *Principia* by substituting area for time in the parabolic measure. And it is that measure of force, $QR/(QT^2 \times SP^2)$, that overshadows his curvature measure, $1/(SY^2 \times PV)$, even in the revised editions of the *Principia* when the latter was introduced as an alternate method. Newton did not, however, acknowledge any debt whatsoever to Hooke.[46]

III. Conclusion

What conclusion is to be drawn from the inspection of these few diagrams? Perhaps, it is that there is more to a diagram than meets the eye. Certainly, the fundamental role played by curvature in the diagram for Lemma 11 is not evident from the figure that accompanies it, nor is the chord of curvature obvious at first glance of the diagram for the revised Proposition 6. Finally, Nauenberg's demonstration of the function of curvature in Newton's diagram of 13 December 1679 underscores its fundamental role in Newton's dynamics, and it provides continuity between Newton's cryptic statement of 1665 and his proposed revisions of 1690.[47] Given that Newton responded to Hooke's 9 December 1679 letter within a few days with the sketch containing some of the information found in the discarded scholium of 1685, it is reasonable to assume that the technique suggested in the cryptic statement of 1665 had been available well before Hooke's challenge of 1679. Moreover, Newton employed curvature in the 1687 *Principia* beyond Book I. Nauenberg pointed out that Newton explicitly applied the curvature measure to the lunar problem in Proposition 28 of Book III.[48] There is evidence, moreover, that curvature was employed by Newton before the publication of the 1687 *Principia.* I have argued, following Whiteside,[49] that Newton employed curvature as a primary measure of force in 1684, when he attempted to reconstruct his 1679 solution for Edmund Halley and that he later communicated this curvature solution to John Locke in 1690.[50] I am convinced that curvature has played an active as well as passive role in Newton's dynamics from its inception in the early 1660s to its formal presentation in the early 1690s. Curvature and its application to Newton's thoughts on dynamics was not just an attempt to provide alternative proofs for the revised *Principia.* It is woven into his conceptual tool kit, just as it is woven into the text of the first edition of the *Principia.* Curvature has its roots in that cryptic statement of 1664, which suggests a solution to elliptical motion using "a tangent circle of equal crookedness;" it continues in all three books of the 1687 *Principia*, and it plays a clear central role in the revised *Principia.* To ignore curvature is to miss an important aspect of Newton's first and continuing thoughts on dynamics: a role manifested by a careful examination of the diagrams.

Notes and References

1. Brackenridge, J. Bruce, *The Key to Newton's Dynamics: The Kepler Problem and the Principia.* Containing an English translation from the Latin by Mary Ann Rossi of Secs. 1, 2 and 3 of Book I, first (1687) edition of Newton's *Principia* (University of California Press, 1995). See Chap. 3 for a discussion of these three methods in the context of Newton's early dynamics before 1669, Chaps. 4 and 5 for the application of the polygonal and parabolic measures after 1679, and Chap. 8 for the demonstration of the circular measure after 1690.
2. The vanishing element of the general curve can be represented by an element of a parabola, where the initial projection velocity is the tangential velocity at the point and the constant acceleration is given in the limit by the approximately constant radial force. No explicit use is made of the properties of a parabola, but the term "parabolic" represents what is implicit in the measure. See Brackenridge,[1] p. 57.
3. Brackenridge, J. Bruce, "The critical role of curvature in Newton's developing dynamics" in *An Investigation of Difficult Things: Essays on Newton and the History of the Exact Sciences* (eds.) P.M. Harman and Alan E. Shapiro (Cambridge University Press, 1992), pp. 231–260. See also Brackenridge[1] Chap. 10 for a discussion of Newton's dynamic measures of force in current mathematical notation.
4. The statement will be given in the following section of this paper.[24] Also see Brackenridge[1] pp. 63–65.
5. M. Nauenberg, "Newton's early computational method for dynamics," *Archives for History of Exact Sciences* **46** (1994): 221–252.
6. See Brackenridge,[1] pp. 157–161 and pp. 202–204 for a discussion of Proposition 9 and Newton's two solutions of the direct spiral/pole problem.
7. Brackenridge,[1] pp. 248–249.
8. Isaac Newton, *The Mathematical Principles of Natural Philosophy*, 3rd Ed. (1726). Translated by I. Bernard Cohen and Anne Whitman, assisted by Julia Budenz, in *Isaac Newton, The Principia* (University of California Press, 1999), pp. 449–450. "These forces tend to the centers of the circles by Proposition 2 and Proposition 1, Corollary 2, and are one to another as the versed sines of the arcs described in minimally small equal times, by Proposition 1 Correspondence 4, that is, as the squares of those arcs divided by the diameters of the circles, by Lemma 7; and therefore since these arcs are as arcs described in any equal times and the diameters are

as their radii, the forces will be as the squares of any arcs described in the same time divided by the radii of the circles. Q.E.D."

For Newton this statement apparently does not need a diagram: By Euclid III, 31 angle ABC between two chords of a circle and a diameter AC is 90° and by Euclid VI, 8 the normal BD from the intersection of the chords B to D on the diameter gives similar triangles ABC and CDB, where DB is the versine. Thus, AB:BC::BC:DB or $BC^2/AB = DB$, where AB goes to the diameter AC as B approaches C, i.e. the versine of the least arc $DB = BC^2/AC$ square of the arc AC applied to the diameter AC.

9. The role of curvature in Lemma 11 was clearly recognized by eighteenth and nineteenth century commentators on the *Principia*. In 1730, just four years after the publication of the third edition, John Clarke commented in some detail upon Newton's qualification "in all curves which have a finite curvature at the point of contact," which was added to Lemma 11 in the revised editions of the *Principia*. Clarke even gave a reference to Milne's *Conic Sections* for a further discussion of curvature: John Clarke, *A Demonstration of Some of the Principal Sections of Sir Isaac Newton's Principles of Natural Philosophy* (London, 1730), p. 89. For a nineteenth century commentator on Lemma 11 and curvature, see Percival Frost, Newton's *Principia, First Book, Secs. I., II., III., with Notes and Illustrations and a Collection of Problems*, 4th Ed. (Macmillan and Co., 1883), pp. 82–113. This excellent pedagogical guide for students preparing for the Mathematical Tripos first appeared in 1878 and was still being published into the twentieth century.
10. However, an English translation of the Latin from the first (1687) edition of Newton's *Principia* by Mary Ann Rossi of Secs. 1, 2 and 3 of Book I is to be found in the appendix of Brackenridge.[1]
11. Brackenridge,[1] p. 252.
12. Newton,[8] p. 69.
13. For a full documentation of both the textual and graphical changes to Proposition 6 in the first three editions of the *Principia*, see Issac Newton, *Mathematical Principles of Natural Philosophy*, 3rd Ed. (1726), with variant readings (eds.) Alexandre Koyré and I. Bernard Cohen (Cambridge University Press, 1972), pp. 103–106.
14. Euclid, *The Thirteen Books of Euclid's Elements*, with introduction and commentary by Sir Thomas L. Heath, 2nd Ed. (Cambridge University Press, 1956), pp. 73–75.

15. The normal YS, which will be required in the revised version of Proposition 10 in the second edition of the *Principia*, appears here in the first edition only in reference to Proposition 16, which does not have a separate figure.
16. *The Correspondence of Isaac Newton*, seven volumes, translated and edited by H.W. Turnbull (Vols. 1–3), J.F. Scott (Vol. 4), A.R. Hall and Laura Tilling (Vols. 5–7), (Cambridge University Press, 1959–1977), Vol. 5, pp. 300–308. The letter of 28 November 1679 from Newton was a response to a solicitation from Hooke, who was then secretary of the Royal Society, for Newton to renew his association with the society. In particular, Hooke asked for Newton's thoughts on Hooke's hypothesis of "compounding the celestial motions of the planets of a direct motion by the tangent & an attractive motion towards the central body." Newton responded that he has not had time recently for "philosophical meditations" nor had he any prior knowledge of Hooke's hypothesis on celestial motion. To "sweeten his reply," as he was later to note, Newton did offer Hooke his thoughts on how to discern the earth's daily rotation from west to east by observing the motion of an object falling freely from a tower near the earth's surface. The common view is that the object would be left behind as it fell and would land to the west of the base of the tower. Newton observed, however, that it would "shoot forward to the east side of the perpendicular," and his figure correctly shows a concave trajectory from the top of the tower to its base. It is a greatly distorted diagram, however, as the object is dropped from a tower whose height appears from Newton's first diagram to be $\frac{2}{3}$ the radius of the earth, yet the gravitational force is assumed to be constant. Even so, all would have been well, had Newton concluded the drawing and text at the point at which the object hits the ground, but he continued the line below the surface of the sphere representing the earth. When the falling body reaches the earth's surface, it continues to move through the earth as if it were a resistive but penetrable medium and the object spirals into the center. In the text of the letter, Newton stated that the object would describe "in it's fall a spiral line," and his curve spiraled into toward the center of the earth. Newton did not say anything more in his text about the spiral line below the surface of the earth, but he did continue to discuss in some detail the initial portion of the curve: the free fall from the top of the tower to the surface of the earth. Nevertheless, the damage was done and Newton's "sweet" soon turned sour.

In his reply of 9 December 1679, Hooke took issue with Newton's spiral. He stated that, "my theory of circular motion makes me suppose it would be very differing and nothing at all akin to a spiral but rather a kind Elleptueid." Hooke's drawing shows that he had (unknowingly?) switched from Newton's rotating frame of reference (in which the tower is at rest) to an inertial frame external to the rotating earth (in which the tower is moving). Thus, the diurnal rotation of the earth is superimposed onto Newton's "spiral". Hooke then continued with two special cases: first, if there were no resistance, then Hooke stated that "the line in which this body would move would resemble an Ellipse," and second, if there were resistance, then Hooke stated that the ellipse would gradually decay and "after many revolutions would terminate in the Center." In addition, Hooke pointed out that in the fall of a ball from a great height (now apparently back in the rotating frame) the deflection "will not be exactly east of the perpendicular but southeast and indeed more to the south than the east."

Newton's response of 13 December 1679 did not take issue with the change of reference frames. Rather, he agreed with Hooke's statement concerning the southeastern deflection for the fall of the ball, but he expressed strong opposition to Hooke's "elleptueid" that was envisioned as the orbit in the absence of resistance. Instead, Newton proposed that for a constant force, the orbit would appear as in the diagram in his letter (Fig. 6), an enlargement of which is shown in Fig. 7.

17. W.W. Rouse Ball, *An Essay on Newton's Principia* (Macmillan and Co., 1893), pp. 138–147.
18. Alexander Koyré, "An unpublished letter of Robert Hooke to Isaac Newton," *Isis* **43** (1952): 312–337, p. 312.
19. Jean Pelseneer, "Une lettre inédite de Newton," *Isis* **12** (1929): 237–254.
20. Johs Lohne, "Hooke vs. Newton: an analysis of the documents in the case on free fall and planetary motion," *Centaurus* **7** (1960): 6–52, p. 27. It is interesting to note that the reproductions of the three diagrams are not printed on the pages of the article, but are separate pieces of paper attached to the page on one side and free on the other three sides.
21. C. Wilson, "The Newtonian achievement in astronomy," in *General History of Astronomy* (eds.) R. Taton and C. Wilson, *Planetary Astronomy from the Renaissance to the Rise of Astrophysics*, Vol. 2, Part A, *Tycho Brahe to Newton* (Cambridge University Press, 1989), p. 243. Nauenberg,[5] p. 222.

22. V.I. Arnol'd, *Huygens & Barrow, Newton & Hooke* (Birkhaüser, 1990), p. 19. This statement may have been based on a flawed replication of the diagram rather than the original. Of the dozen or more articles published on the topic before 1990, only two contained an actual reproduction of the original drawing, and all of the replicas were flawed.
23. Nauenberg.[5]
24. John Herivel, *The Background to Newton's Principia* (Oxford, 1965), p. 130. There are two subscription errors in this first published version of this statement: "If the body b moved in an Ellipsis *that* [instead of then] its force in each point (if its motion in that point be given) *will* [instead of may] be found by a tangent circle of Equal crookedness with that point of the Ellipsis." See Brackenridge,[1] p. 282, note 11, for a discussion of this transcription.
25. Issac Newton, *The Mathematical Papers of Isaac Newton*, eight volumes. Translated and edited by D.T. Whiteside (Cambridge University Press, 1967–1981). See Vol. 6, pp. 538–559 for the original published account of these proposed revisions (following 1690), which includes an English translation from the Latin and extensive commentaries. Of the many valuable notes by Whiteside, see particularly note 25, pp. 548–550 for a discussion of role curvature played in these revisions, which were eventually published in the revised *Principia* as alternate solutions. See also Brackenridge,[1] pp. 166–210 for a discussion of Newton's unpublished and published revisions and extensions of this material.
26. See Ref. 38 for a list of published replicas of Newton's 13 December 1679 diagram.
27. Pelseneer,[19] p. 250. An integral gives the angle θ between the maximum radius R and minimum radius ρ (i.e. between apogee and perigee), with the result that the maximum angle for a constant force is given by $180°/\sqrt{3} = 103°55'$. See note 29 below for an extended discussion of this integral.
28. Lohne[20] developed an approximate expression for some of the analysis underlying the integral for the angle given earlier by Pelseneer.[19] It is important to note that the analysis calls upon the law of areas, which was not available to Newton before his 13 December 1679 letter and diagram. Lohne also offered three different approaches: the first using elemental parabolic arc, the second using tangents, and the third using curvature. It is this third measure that he used to obtain what he called the "correct curve," but he gave no indication of how it was done other than the

following statement: "We can compute the radii of curvature from the relationship

$$v^2/\rho = g_n \quad \text{or} \quad \rho = v^2/g\cos\beta$$

where g_n is the component of g normal to the curve. This last method gave us Fig. 8 (Lohne's 'correct curve')," p. 44. However, although Lohne employed curvature in this measure, he was not able to relate his results to the curve Newton had in his 13 December diagram. Lohne concluded that a comparison of curves indicated that Newton "had greatly overestimated the precession of the auges." p. 45.

29. D.T. Whiteside, "Newton's early thoughts on planetary motion: a fresh look," *British Journal for the History of Science* **2**: (1964) 117–137, p. 134, note 55. In this note, Whiteside supplied an alternative to the equation given by Pelseneer in 1929 for the angle φ between apogee and perigee. In 1960, Lohne[20] developed an expression for some of the analysis underlying the integral for the angle given by Pelseneer in 1929.[19] It is important to note that the analysis calls upon the law of equal areas (i.e. modern conservation of momentum) which was not available to Newton before 1679, and thus before his 13 December 1679 letter and diagram. The analysis also employs the velocity relationship for a constant acceleration g, which may be obtained from the basic relationship $vdv/dr = -g$, and thus $v^2 = v_0^2 + 2g(R-r)$, (i.e. modern conservation of mechanical energy). Lohne generated the following differential equation for φ:

$$\frac{d\varphi}{dr} = \frac{Rv_0}{\sqrt{(R-r)[2gr^2 - v_0^2(R+r)]}}$$

He obtained an approximate value of $gR/3$ for v_0 by estimating from Newton's diagram that the minimum radius is $R/2$. Whiteside supplied the equation given by Pelseneer in 1929 for the angle φ between apogee and perigee, which can be obtained by substituting the exact expression for the initial velocity into Lohne's final analytical result. In contrast to Lohne's approximate value for v_0, Whiteside provided an alternate expression by setting $\frac{dr}{d\varphi} = 0$ at $r = \rho$, where ρ is the minimum radius at perigee. When this value is substituted into the differential equation above, it can be written as follows:

$$\frac{d\varphi}{dr} = \frac{R\rho}{r\sqrt{(R-r)(R-\rho)[r(R+\rho)R\rho]}}$$

If a new variable $\sin\theta = [r(R+\rho) - 2R\rho]/r(R-\rho)$ is defined, where $\lambda = (R - \rho/(R+\rho)$, then the equation can be written in integral form, with the approximate value given by Whiteside as follows:

$$\varphi = \int_{-\pi/2}^{\pi/2} [(1+\lambda\sin\theta)/(3+\lambda\sin\theta)]^{1/2} d\theta$$

$$\cong 180°/[3 + (1/2)\lambda^2 + (1/2)\lambda^4]^{1/2}$$

This approximation holds only to second order in λ but it is exact for $\lambda = 0$ and $\lambda = 1$. Thus, φ is a maximum of $180°/\sqrt{3}$ (i.e. 103°55) for $\lambda = 0$ or $R = \rho$, that is, for a nearly circular orbit, and has a minimum of 180°/2 (i.e. 90° for $\varphi = 0$ or $\lambda = 1$), that is, for a nearly linear orbit. It is quite clear however, that Whiteside did not attribute this analysis, or anything similar, to Newton at the time of the 13 December 1679 letter and diagram. He referred to it as "a 'Newtonian' analysis," with the clear implication that it is not only after the area law, but that it is a modern analytical expression. Whiteside concluded (correctly) "but it is clear that Newton at this time [1679] had no such exact theory."

30. Herivel,[24] p. 244, note 1. "It is chiefly memorable for the evidence it provides of the undeveloped state of Newton's thought on the problem at what was presumably only a short time before he arrived at a definite solution. There is, for example, no firm indication that he had actually attempted a quantitative solution.... However, Newton's assertion that he might have added something to the problem 'by points *quam proxime*' could be interpreted as an indication that his thought on the problem had advanced further than he was prepared to admit to Hooke." Herivel's replication of the diagram that accompanies the text (p. 243) is closed and does not touch the outer circle of apogee at the point A.
31. R.S. Westfall, *Force in Newton's Dynamics* (American Elsevier, 1971), p. 428–429 "Despite the fact that he treated gravity as constant in the letter, Newton revealed his command of a rudimentary, relatively systematized quantitative dynamics.... The analysis as given can be faulted.... Whatever its faults, however, it was pregnant with future possibilities." Westfall also discussed this diagram in his later work, R.S. Westfall, *Never at Rest: A Biography of Isaac Newton* (Cambridge University Press, 1980), p. 386. The diagrams in both works are identical and both show the curve incorrectly as a closed figure.

32. The Latin text of this scholium was first published in 1966, but without diagram, comment, or translation, by J.W. Herivel,[24] p. 325.
33. This diagram is not an attempt to replicate Newton's original curve, rather it is generated from Whiteside's analytical solution discussed in note 29 above.
34. Newton[25] VI, pp. 369–382. The final draft of Proposition 45 with Whiteside's notes.
35. Wilson[21]. The comments on Newton's letter of 13 December to Hooke occupy just a small portion of this excellent chapter (pp. 242–243). Wilson's major concern was with when and how Newton first came to accept the universality of the law of gravitation, and he concluded that "the first clear indication that Newton is thinking of gravitation as universal" was not before December 1684, and appeared in the augmented version of *De motu corporum* (p. 253). See also Wilson's discussion of Newton's attempt to calculate the motion of the Moon's apse (pp. 262–263), and Nauenberg's article in this volume which resolved the difficulty that Wilson encountered.
36. Nauenberg.[5]
37. Koyré,[18] p. 331 and Turnbull,[16] II, p. 307.
38. Mirror symmetry is missing in *all* the attempts to replicate Newton's original diagrams: both in the publications up to the last decade, i.e. Pelseneer (1929),[19] Koyré (1952),[18] Turnbull (1960),[16] Whiteside (1964),[25] Herivel (1965),[24] Westfall (1971/1980),[31] and in those of the last decade, i.e. P.J. Pugliese, "Robert Hooke and the dynamics of motion in a curved path," *Robert Hooke, New Studies* (eds.) M. Hunter and S. Schaffer (Boydell Press, 1989), pp. 181–298; Julian B. Barbour, *Absolute or Relative Motion? A Study from a Machian Point of View of the Discovery and the Structure of Dynamical Theories* (Cambridge University Press, 1989), p. 544; Wilson (1989);[21] H. Erlichson, "Newton's 1679/1680 solution of the constant gravity problem," *American Journal of Physics* **59** (1990): 728–733; A. Rupert Hall, *Isaac Newton Adventurer of Thought* (Cambridge University Press, 1992), p. 203; and F. De Gandt, *Force and Geometry in Newton's Principia*, translated by C. Wilson (Princeton University Press, 1995). Mirror symmetry is a consequence of the exact dynamics and therefore is found in the diagrams generated by numerical integration of the analysis in Lohne's 1960 paper[20] and in the diagram Whiteside generated in 1974[25] for the deleted scholium, neither of which is an attempt to replicate Newton's original flawed diagram.

39. Nauenberg,[5] p. 237.
40. Nauenberg,[5] p. 226.
41. Nauenberg,[5] p. 237.
42. The velocity can be obtained from the relationship $dv/dt = vdv/dr = A$ (the constant acceleration), which when integrated gives $v^2 = 2(E - Ar)$, where E is the constant of integration given from the initial conditions. The new tangential component of velocity $v_{(t)}$ can also be approximated by the simple iteration technique of $v_{(t)n+1} = v_{(t)n} + \delta v_{(t)n} = v_{(t)n} + a_n \cos\theta\delta t_n$, where in this case a_n is a constant, δt_n is given by $\delta = 22° = \omega\delta t_n = (v^2_{(t)n}/\rho_n)\delta t_n$, and I simply measured the angle θ between the tangent and the radius directed toward the center of force C with a protractor. Thus, the new radius of curvature ρ_{n+1} is given by $(v^2_{(t)n+1})/(a \sin\theta)$.
43. Westfall,[31] p. 430 (1971).
44. For example, Newton revised the discussion of circular motion in the scholium to Proposition 4 in the first edition of the *Principia* by the insertion of the word "centrifugal" into the revised second edition. Koyré and Cohen,[13] p. 101. See also Bertoloni Meli, Domenico. Equivalence and Priority: Newton vs. Leibniz (Oxford, 1993), pp. 177–185 for an extended discussion of this topic.
45. Nauenberg,[5] p. 244. The distinction that has traditionally been made between *centrifugal* (center-fleeing) and *centripetal* (center-seeking) is simply that the former is outward and the latter is inward (see for example, Westfall,[30] 1971, p. 427: "Hooke was the one who in fact set upright the crucial problem of orbital motion, which had been conceptualized, as it were, upside down.) It is true that in his early discussion of circular motion, Newton, following Huygens, employed the word *centrifugal* in his discussion of circular motion. Moreover, in that early usage the word represented, following Descartes, an outward endeavor of the body to recede along the normal to the circle. When Newton made his calculation of circular motion however, it was the implicit unnamed inward reaction to the *centrifugal* that was employed, and the only difference between the two forces for uniform circular motion was that one was outward and the other was inward.

    Newton's curvature analysis of general orbital motion in the 13 December 1679 letter to Hooke still employs the technique used by Newton in his analysis of uniform circular motion to piece together the general orbital motion, "and it is *only* in the context of uniform circular motion (with the exception of a repulsive force in Proposition 12) that Newton employs the word *centrifugal.* In this analysis Newton" resolves the motion into tangential and *normal* components, neither of which is directed toward the center of force. Hooke's suggestion is that one looks instead toward the tangential and *radial* components of motion (that produced by the tangential velocity and that produced by the change in velocity due to the impulsive force) where the radial component is directed toward the center of force; that is, the center seeking or *centripetal* force. It is from this "continuous normal/ impulsive radial" distinction, and not from a simple opposition to the *centrifugal* force, that Newton was led to coin the word *centripetal.*

46. Nauenberg,[5] p. 244. "The missing ingredient for a complete solution... was provided by the fundamental idea of Hooke to view orbital motion as compounded by the tangential inertial velocity and a change of velocity impressed by the central force. This idea can be expressed in simple mathematical form for forces which act in short pulses, for which the curvature method is not applicable.... After the correspondence with Hooke, Newton evidently understood the full equivalence, for vanishingly small time steps, of these two distinct physical approaches to orbital motion, but never credited Hooke for his seminal contribution."
47. Nauenberg.[5]
48. Newton,[8] p. 267. Book III, Proposition 28, Problem 9 "*To find the diameters of the orbit in which the moon would have to move, if there were no eccentricity.*" The curvature of the trajectory that a moving body describes, if it is attracted in a direction which is everywhere perpendicular to that trajectory, is as the attraction directly and the square of the velocity inversely. I reckon the curvatures of lines as being among themselves in the ultimate ratio of the sines or of the tangents of the angles of contact, with respect to equal radii, when those radii are diminished indefinitely.... Newton,[13] p. 624 notes that this text is unchanged from the 1687 edition. This reference to curvature in the first edition was called to my attention by Michael Nauenberg.
49. *The Preliminary Manuscripts for Isaac Newton's 1687 Principia 1684–1685* (ed.) D.T. Whiteside (Cambridge University Press, 1989). See Whiteside's note 53 on pages xx–xxi.
50. J. Bruce Brackenridge, "The Locke/Newton manuscript: conjugates, curvatures, and conjectures." *Archives internationales d'histoire des sciences*, **43** (1993): 280–292.

George E. Smith

Fluid Resistance: Why Did Newton Change His Mind?

GEORGE E. SMITH

Philosophy Department, Tufts University
Medford MA 02155, U.S.A.

1 Introduction

Section 7 of Book II was most extensively and substantially rewritten in the second edition of the *Principia*. There, Newton was trying to reach conclusions about the magnitude of resistance forces and the contribution made to these forces by the inertia of the fluid medium. The issue of the magnitude of the forces, once Newton had chosen a dimensionless way of representing them, boiled down to finding the magnitude of a generic coefficient, at least for spheres. This he did experimentally, employing entirely different experiments in the second edition from the ones he had used for this purpose in the first, and obtaining a distinctly different magnitude. Newton separated the issue of the contribution to resistance made by the inertia of the fluid into two parts: (1) the force the inertia of the fluid imposes on the front of a moving body, and (2) the force, positive or negative, resulting from any difference that motion induces in the pressure of the fluid acting on the rear of the body. He devised theoretical models for determining the inertial resistance on the front of the body and then did what he could to surmise the induced force on the rear. The conclusions he reached on all three of these points — the dimensionless magnitude of the force, the inertial component on the front, and the induced component on the rear — changed substantially from the first edition to the second.

In order to indicate the extent of these changes, I need to adopt a way of non-dimensionalizing resistance forces. Newton did not settle on his preferred way of non-dimensionalizing them until the second edition.[1] In spite of the anachronism, I find it easier here to use our modern non-dimensionalization, employing the *drag coefficient*:

$$C_D = \frac{2F_{\text{RESIST}}}{\rho_f A_{\text{front}} v^2}$$

Table 1 Fluid resistance: how Newton changed his mind.

	1st Edition	2nd Edition
Non-dimensional inertial resistance force on the front face of a sphere in a "rarified" fluid (theory)	*$C_D = 2.0$	$C_D = 2.0$
Non-dimensional inertial resistance on the front face of a sphere in a "continuous" fluid (theory)	$C_D = 2.0$	$C_D = 0.5$
Non-dimensional measured total resistance force on a sphere	$C_D \approx 0.7$–0.9	$C_D \approx 0.5$
Ratio of the fluid force induced on the rear face of a moving sphere to the inertial resistance force on the front face	$\frac{2}{3}$	0

*Where C_D is the drag coefficient and rarified fluid particles are perfectly elastic.

where ρ_f is the density of the fluid medium, A_{front} is the frontal area of the body, v is the relative velocity between the fluid and the body, and F_{RESIST} is the fluid resistance force. While Newton never employed this non-dimensionalization, he was responsible for introducing the term forming the denominator.[2]

Table 1 summarizes the changes Newton made between the first and second editions. In the first edition, he concluded from theoretical models of different types of fluids that fluid inertia produces a resistance force on the front of a sphere that amounts to a drag coefficient of 2.0 — whether the fluid be "rarified" (with particles perfectly elastic) or "continuous". From pendulum-decay experiments, he obtained a magnitude for the total resistance force on a sphere that amounts to a drag coefficient more or less in the range from 0.7 to 0.9, and a magnitude for the inertial component of this force that amounts to a drag coefficient in the range from 0.65 to 0.85, with the uncertainty stemming from vagaries in his data. Based on these findings, he concluded that the force induced by the action of the fluid on the rear face of the sphere is substantial in both air and water, counteracting as much as $\frac{2}{3}$ of the inertial resistance force on its front face.

In the second edition, Newton obtained the same result from his theoretical model of rarified fluids, but he lowered the inertial resistance on the front face of a sphere moving in a continuous fluid by a factor of four — in effect, dropping the drag coefficient from 2.0 to 0.5. In making this change he also decided, in contrast to the first edition, that the force induced on the rear face is negligible in continuous, as well as in rarified, fluids. Finally, he concluded from vertical-fall experiments that the resistance on spheres, from fluid inertia and all other effects, amounts to a drag coefficient very near to 0.5 in both air and water, with increments above this value due to such non-inertial mechanisms as the lack of lubricity of the fluid. The question is, what led Newton to change his mind about the magnitude and make-up of fluid resistance forces in the quarter century between the first and second editions?

2 Newton's Problem of Fluid Resistance

In Books I and III, Newton successfully "deduced" the centripetal forces acting on celestial bodies from phenomena of celestial motion. Similarly in Book II, he attempted to deduce the forces of resistance acting on bodies from phenomena of motion in fluid media. In the case of celestial forces, the mathematical framework he adopted as a working hypothesis was that the forces are directed toward a center, and their strength is a function of distance from it, $f(r)$. The analogous mathematical framework he adopted as a working hypothesis in the case of resistance forces was that they are directed oppositely to the motion, and their strength is an additive function of powers of the relative speed v between the body and the fluid, most likely the function[3]

$$F_{\mathrm{RESIST}} = a_0 + a_1 v + a_2 v^2 \tag{1}$$

His problem then was to infer laws governing the magnitude and variation of a_0, a_1 and a_2 from phenomena of motion in fluids in analogy with the way in which the law of universal gravity, $F_{\mathrm{CENTRAL}} = Gm_1m_2/r^2$, was inferred from phenomena of celestial motion.

Newton had no *a priori* basis for saying anything about a_0 or a_1,[4] but he did conclude from the outset that a_2 should vary as the product of the density of the fluid and the frontal area of the moving body. He seems to have reached this conclusion by thinking of the fluid as consisting of non-interacting particles, like debris scattered in space. Resistance forces arise from the moving body impacting these particles. The force per impact will vary as the mass of the particles and the impact velocity, and the number of impacts will vary as the velocity and the product of the frontal area and the number of particles per unit volume. Although this reasoning does not apply directly to what Newton called "continuous" fluids, he concluded that a_2 varies with fluid density and frontal area in the same way for them, with the full term representing the inertial resistance of the fluid.

His refined working hypothesis then, was that resistance forces can be represented, at least to a first approximation, within the following mathematical framework:

$$F_{\text{RESIST}} = a_0 + a_1 v + b_2 \rho_f A_{\text{front}} v^2 \tag{2}$$

where b_2 may vary with the shape of the body. For spheres (2) can be rewritten as:

$$F_{\text{RESIST}} = a_0 + a_1 v + c_2 \rho_f d^2 v^2 \tag{3}$$

where d is the diameter and c_2 was expected to be a constant. The problem Newton faced was first to confirm the $\rho_f d^2$ constituents of the v^2 term and then to characterize a_0, a_1 and c_2, if not b_2, from phenomena of motion.

In order to draw conclusions from phenomena of motion, Newton first needed mathematical solutions for motions. He could then work backwards from measurements of time and distance to the magnitudes of the components of the resistance force. Much of Book II is devoted to mathematical solutions for problems of motion under various types of resistance. Section 1 gives solutions for horizontal, vertical, and projectile motion under uniform gravity with resistance varying as v. Section 2 gives corresponding solutions for horizontal and vertical (but not for projectile) motion with resistance varying as v^2.[5] Section 3 combines the solutions for v and v^2. Section 4 solves the orbital-decay problem for circular orbits under inverse-square gravity with resistance varying as v^2. Section 6 offers solutions for pendulum motion with resistance varying as a linear combination of terms in v to different powers.

As Tom Whiteside has emphasized, Newton's pendulum-decay solution extends to any power of v, integer or otherwise.[6] Hence, pendulum-decay motion in principle gave Newton an avenue for inferring forces of resistance that freed him from the simple mathematical framework I gave earlier, allowing these forces instead to result from any linear combination of effects proportional to powers of v. In this regard, Sec. 6 provides flexibility in attacking the question of resistance forces of the same sort that Secs. 8 and 9 of Book I provide in attacking the question of celestial central forces.

Section 7 comes after these solutions for motion in resisting media. In it, Newton turned to considering ways in which macrophysical resistance forces might arise from microstructural effects. In other words, Sec. 7 of Book II parallels Secs. 12 and 13 of Book I, where Newton considered ways in which central forces might arise from forces among particles forming bodies. Up to the end of Sec. 6, proper — i.e. Sec. 6 without the General Scholium that Newton shifted to the end of it in the second edition — the changes in Book II from the first edition to the second are comparatively minor. From there on, however, we need to look at the individual editions.

3 Resistance Forces in the First Edition

Newton gave a microstructural account of the component of resistance arising from the inertia of the fluid — i.e. the component that varies as $\rho_f v^2$ — for two *theoretically defined* types of fluid. A *rarified* fluid consists of particles spread out in space; a *continuous* fluid consists of particles so packed together that each is in contact with its neighbors.[7]

3.1 *The first edition — rarified fluids*

The microstructural mechanism producing inertial resistance in a rarified fluid is motion transferred from the moving body to the individual fluid particles with which it collides, as if the body were moving through scattered debris in empty space. Newton used a non-dimensionalization for this inertial resistance in the first edition that he abandoned in subsequent editions, and he added the qualifier that the result holds only approximately.[8] In modern terms, his result for a sphere is equivalent to a drag coefficient C_D of 2.0 if the particles are maximally reflective, 1.0 if they do not reflect at all, and between these values otherwise. This result is based purely on the impact between the particles and the forward face of the body. It presupposes that no action is occurring among the particles, altering the fluid ahead of the body or, for that matter, behind or to the sides of it. As a consequence, the fluid at the rear of a moving body is irrelevant to the inertial resistance force in a rarified fluid.

A corollary to the result for rarified fluids that appears only in the first edition indicates that the actual resistance will be somewhat greater, especially at slow speeds, owing to non-inertial effects. A further corollary then suggests an experimental program: first determine the extent of the elastic rebound, and hence the rule giving the magnitude of the inertial resistance, by experiments at high speeds; and then turn to the difference between the magnitudes given by this rule and observed resistance at slower speeds in order to obtain a rule for the non-inertial resistance in rarified fluids.[9]

Newton also concluded that inertial resistance in rarified fluids varies with the shape of the body as well as its frontal area. His celebrated result on the solid of least resistance is for rarified fluids only. While the mathematics of this result is noteworthy,[10] how much importance Newton attached to his theory of rarified fluids is open to question. In his comparisons of theoretical calculations with vertical-fall experiments in the second and third editions, he expressly treated both water and air as continuous fluids.[11] Whether he held this view in the first edition is not so clear. In the pendulum-decay experiments in both editions, he expected the fluid resistance in water versus air to match their respective densities. In the first edition, however, this may reflect nothing more than his thinking that the non-dimensionalized resistance is the same in rarified and continuous fluids, so that this distinction is irrelevant when comparing theoretical and measured values. As the non-dimensionalized resistance is no longer the same for the two theoretical types in the second edition, his lumping water and air together in pendulum-decay there (and in the third edition) by implication relegates rarified fluids to a purely hypothetical status. At least in the second and third editions, then, if not the first, rarified fluids drop from sight when Newton turned to experimental results.

3.2 *The first edition — continuous fluids*

Continuous fluids posed a more difficult problem for Newton. He treated inertial resistance in these fluids as arising predominately from the moving body's having to push a column of fluid in front of it, or with the body still and the fluid moving, as the force of a column of fluid impinging like a weight on the object. Picture water passing around the sphere P in the cylindrical channel in Fig. 1, taken from the first edition. Part of the fluid forms a "stagnation" column above the sphere, its weight supported by the sphere. Recall that Newton's problem, in effect, was to derive the value of c_2 in the term $c_2 \rho_f d^2 v^2$. His method was to determine the weight of the water supported by the sphere and the velocity of the water flowing by it, both versus the height of the water above the hole in Fig. 1. Given this weight (i.e. the inertial fluid force on the front face, $c_2 \rho_f d^2 v^2$) and the velocity, he could infer the (dimensionless) value of c_2 for the front face.

In the first edition, Newton concluded that the weight supported by the sphere is that of the entire cylindrical column of fluid of its diameter above it. He further concluded, mistakenly, that the fluid effluxing through the orifice N will rise to half the height of the fluid in ABCD, so that the velocity of the fluid

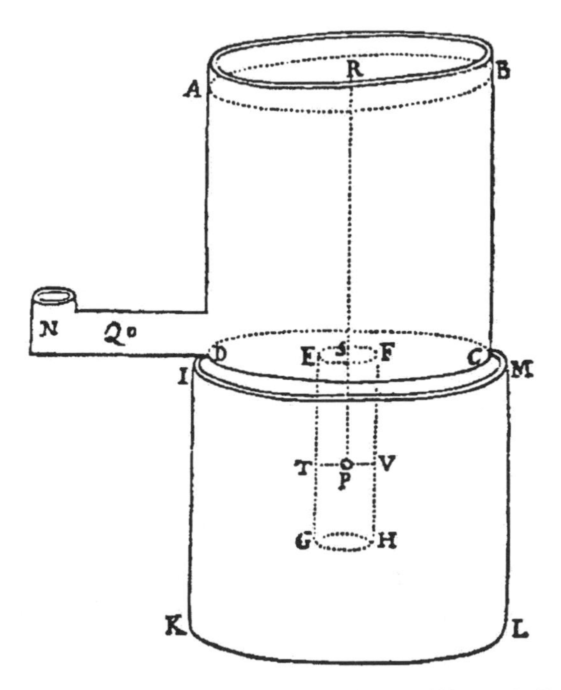

Fig. 1 The efflux problem diagram from the first edition of Newton's *Principia*. The two efflux streams contain spheres P and Q on which the fluid exerts forces.

passing the sphere P, or equally the sphere Q, is that needed to reach half this height.[12] (Newton was probably led to this mistaken conclusion not by the reasoning given in the *Principia*, but by measuring the volume of water that effluxes from a hole in the bottom of a container in a given time and inferring the velocity from the diameter of the hole.)

Putting this velocity and the weight of the column together, the conclusion reached in Proposition 38 was that the inertial resistance on the front of a sphere in continuous fluids amounts, in modern terms, to a drag coefficient of 2.0.[13] In the first corollary of this proposition, Newton remarked that this magnitude is the total inertial force if the sphere "while moving is urged from the rear by the same force as when it is at rest.... But if while moving it is urged from behind less, it will be more retarded; and conversely, if it is urged more, it will be less retarded."[14] The remaining corollaries and the last two propositions of Sec. 7 argue, against Descartes, that "if fluids, however subtle, are dense, their force to move and resist solids is very great."[15] In the third corollary, Newton remarked that media exerting little resistance on moving bodies must be

> "not only very fluid but also far rarer than those bodies that move in them, unless perhaps someone should say that every very fluid medium, by a perpetual impulse, made upon the back part of a projectile, promotes its motion as much as it impedes and resists on the front part. And indeed it is likely that some part of that motion that the projectile impressed upon the medium is given back to the body from behind by the medium carried circularly. For also having made certain experiments, I have found that in sufficiently compressed fluids some part is given back. But that it is all given back in any case whatsoever neither seems reasonable nor squares well with the experiments that I have tried hitherto."[16]

3.3 *The first edition — measured forces*

Appended to Sec. 7 in the first edition is a General Scholium giving the results of pendulum-decay experiments and drawing conclusions from them. Resistance forces, as Newton conceived them, pose a special problem when trying to reach conclusions from experiments: suppose that the total force is the sum of two or more terms in different powers of v, representing different physical mechanisms such as the viscosity and inertia of the medium; then experiments must *disaggregate* the mechanisms in order to yield conclusions about the individual terms. Newton saw pendulum-decay as a way to accomplish this.

The solution for pendulum motion in Sec. 6 gives a systematic relationship between the amount of arc lost per swing, non-dimensionalized with respect to the length of the pendulum, and the ratio of the maximum resistance force on the bob to its weight. Thus, if part of the lost arc is proportional to a resistance-as-v effect, part proportional to a resistance-as-v^2 effect, etc., the total lost arc in a single swing can be represented as a sum:

$$\delta\,\text{arc} = A_0 + A_1 V_{\text{max}} + A_2 V_{\text{max}}^2 + \cdots + A_n V_{\text{max}}^n \tag{4}$$

where the $A_i s$ are constants for a given pendulum. Under the assumption that the resistance is weak enough for the ratio between the velocity everywhere along the arc and the maximum velocity to remain virtually the same as in the unresisted case, the solution for pendulum motion allows the resistance force at the point of maximum velocity to be represented by a corresponding sum:

$$F_{\text{RESIST}}/\text{WEIGHT} = [(1/2)A_0 + (2/\pi)A_1 V_{\max} + (3/4)A_2 V_{\max}^2 + \cdots](1/\ell) \quad (5)$$

And in general, as Tom Whiteside had shown, if part of the lost arc is proportional to v^n, then, whether n is an integer or not, the corresponding contribution to the total resistance given by Newton's solution amounts to:[17]

$$\frac{1}{2\int_0^1 (1-x^2)^{n/2} dx} A_n V_{\max}^n$$

Now, $V_{\max}$ in a cycloidal pendulum, without resistance, is proportional to the arc length. Hence, by starting the pendulum at different points and measuring the average arc lost per swing in each case, the values of the A_i can be inferred via simultaneous algebraic equations: for example,

$$\begin{aligned} \delta\,\text{arc}_4 &= A_0 + A_1 V_{\max} + A_2 V_{\max}^2 && \text{for a four-inch arc} \\ \delta\,\text{arc}_{16} &= A_0 + 4A_1 V_{\max} + 16A_2 V_{\max}^2 && \text{for a 16-inch arc} \quad (6) \\ \delta\,\text{arc}_{64} &= A_0 + 16A_1 V_{\max} + 256A_2 V_{\max}^2 && \text{for a 64-inch arc} \end{aligned}$$

where $V_{\max}$ is now the maximum velocity for a four-inch arc. Thus, so long as the resistance does not significantly alter the velocity ratios along the arcs, Newton had a mathematical solution for the resisted motion of pendulums. This solution offered the promise of disaggregating the effects contributing to resistance, allowing him to infer values of the individual A_i for a given pendulum, and hence the magnitude of each component of the resistance force, from the amounts of arc lost by this pendulum.

Newton employed circular rather than cycloidal pendulums in his actual experiments, arguing that the two principal deviations from the cycloidal theory introduced thereby essentially cancel one another. His experiments fall into four groups. In the primary group, he used a 126-inch long pendulum with a $6\frac{7}{8}$-inch diameter spherical wooden bob to obtain basic values for the resistance forces. He next changed to a two-inch diameter lead bob in order to determine the effect of the size of the bob. In the third group, he used a $134\frac{3}{8}$-inch long pendulum to compare resistance in air and water, with the bob moving in a specially constructed trough for purposes of the latter. Finally, he used a much shorter pendulum to compare resistance in water and mercury.

Unfortunately, the data from these experiments, displayed in part in Table 2, were disappointing. One shortcoming was their failure to yield stable values for the coefficients of terms in powers of velocity less than two.[18] Newton tried at least four combinations of terms in powers of V less than two before selecting terms in V and $V^{3/2}$.[19] In all the combinations, the coefficients of the terms in powers of V less than two vary by anywhere from 200 to 600%, depending on which three data points are used to determine them. Worse,

Table 2 Newton's data from his primary pendulum-decay experiments in air.

Arc (in)	Lost Arc (in)	Number of Oscillations	Average δ_{arc} (in)	Relative v_{max}	Consecutive Ratios
2	$\frac{1}{4}$	164	$\frac{1}{656}$ (1.524E-3)	$\frac{1}{2}$	2.71
4	$\frac{1}{2}$	121	$\frac{1}{242}$ (4.132E-3)	1	3.51
8	1	69	$\frac{1}{69}$ (1.449E-2)	2	3.89
16	2	35.5	$\frac{4}{71}$ (5.634E-2)	4	3.83
32	4	18.5	$\frac{8}{37}$ (2.162E-1)	8	3.82
64	8	9.67	$\frac{24}{29}$ (8.276E-1)	16	
2	$\frac{1}{2}$	374	$\frac{1}{748}$ (1.337E-3)	$\frac{1}{2}$	2.75
4	1	272	$\frac{1}{272}$ (3.677E-3)	1	3.34
8	2	162.5	$\frac{4}{325}$ (1.231E-2)	2	3.90
16	4	83.3	$\frac{12}{250}$ (4.800E-2)	4	4.00
32	8	41.67	$\frac{24}{125}$ (1.920E-1)	8	3.68
64	16	22.67	$\frac{48}{68}$ (7.059E-1)	16	

In the first of the two sets of data, Newton counted the number of oscillations until $\frac{1}{8}$ of the total initial pendulum arc was lost; in the second set, he counted the number until $\frac{1}{4}$ of the initial arc was lost. He then divided to find the average arc lost per oscillation, δ_{arc}. The column labeled "Relative v_{max}" gives the ratio of the maximum velocities, ignoring resistance, to the maximum velocity with a four-inch initial arc. The column labeled "Consecutive Ratios" gives the ratio of the average arcs lost per oscillation between consecutive rows; Newton expected the numbers in this column, in the absence of vagaries in the data, to be monotonically increasing, approaching the square of the ratios of the corresponding values of v_{max}, i.e. 4.0, as initial arc length was increased. This column thus serves two purposes: (1) it gives grounds for concluding that no power in velocity greater than two is entering into the resistance forces; and (2) it gives indications of vagaries in the data.

for many data points these coefficients turn out negative. Newton gave no reason for the selection he used in the text, nor did he offer any thoughts on the physics lying behind the $V^{3/2}$ term. Most of the data combinations imply a negative value for this coefficient. The one redeeming feature of it that I was able to find is that, with it, the values of the coefficients of both the V and V^2 terms lie near the middle of the values obtained from different data combinations and different assumptions about the third term. Regardless, the pendulum-decay experiments shed no light at all on how resistance terms in powers of velocity less than two vary with such factors as the size of the body and the properties of the fluid.

The results were somewhat better for the coefficient of the V^2 term, but its value too was not entirely stable. For example, when the $V^{3/2}$ term is included, the coefficient of the V^2 term, as Newton calculated it,[20] changes by 13% between the two data sets shown in Table 2, the former set using the number of swings before $\frac{1}{8}$ of the initial arc is lost to determine the average arc lost per oscillation, and the latter using the number before $\frac{1}{4}$ is lost. Even so, the results still made clear that the V^2 term totally dominates the others, and thereby masks them, in both air and water. The magnitude of the V^2 term failed to vary in proportion with d^2 when Newton changed the diameter of the bob or with ρ_f when results in air, water and mercury were compared.[21] Newton argued that these failures resulted from fluid resistance acting on the pendulum string, which was ignored in the theoretical solution for pendulum motion. After introducing empirically-based corrections for this effect, Newton offered a reasonable argument that the v^2 resistance term for spheres does vary as $\rho_f d^2$. Among other things, this supported the conclusion that this term represents the inertial action of the medium.[22]

3.4 *The first edition — comparing theory and experiment*

Near the end of the General Scholium, Newton turned to the question of how the inferred magnitude of the v^2 term compares with the theoretical values he had obtained in Sec. 7.[23] An analysis of his data reveals a disaggregated component of inertial resistance equivalent to a drag coefficient more or less in the range from 0.65 to 0.85, between roughly $\frac{1}{3}$ and less than $\frac{1}{2}$ of his theoretical value. Newton himself said in the first edition that the magnitude of the inertial component implied by the data was around $\frac{1}{3}$ of the value given in Proposition 38.[24] Based on this, he concluded that a counteracting inertial force is induced by the action of the fluid on the rear face of the moving body, canceling roughly $\frac{2}{3}$ of the inertial resistance on the front face.[25] (This presumably was the experimental result he alluded to in the third corollary to Proposition 38, quoted earlier.) Save for the vagaries in the pendulum-decay data, Newton had no obvious reason to doubt any of these conclusions when the first edition appeared.

4 The Years Between the First and Second Editions

Newton ended up rejecting most of these conclusions in the years between the first and second editions. He rejected his theoretical value for the inertial resistance on the front of a sphere in a continuous fluid, reducing the dimensionless magnitude of this force by a factor of four. He rejected his experimentally determined measure of the inertial resistance on a sphere, concluding that the pendulum-decay experiments had given a value 35% or more too high. And he rejected the idea that the action of the fluid on the rear face of the sphere is significant, concluding instead that it is at most a second order effect. What convinced him to do this?

Part of the answer is well-known: he decided, under the urging of Fatio de Duillier, that his solution for the efflux velocity was wrong; and he carried out some vertical-fall experiments in a trough of water, getting results in conflict with the pendulum-decay experiments. As we shall see, however, these two are not enough to explain all the changes.

4.1 *The revised efflux solution*

From Volume 3 of the *Correspondence*, we learn of a note Fatio had inserted in the margin of his copy of the *Principia*, beside the efflux solution: "I could scarcely free our friend Newton from this mistake, and that only after making the experiment with the help of a vessel which I took care to have provided."[26] The efflux velocity is that acquired in falling not half the height, but instead the full height of the fluid in the container. Years later, Newton cited two experiments: one in a letter to Cotes and the other in the second edition. In the letter, he stated:

> "For I have seen this experiment tried & it has been tried also before the Royal Society, that a vessel a foot & an half or two foot high & six or eight inches wide with a hollow place in the side next the bottom & a small hole in the upper side of the hollow, being filled with water; the water which spouted out of the small hole, rose up in a small stream as

high as the top of the water which stagnated in the vessel, abating only about half an inch by reason of the resistance of the air."[27]

In the second edition of the *Principia*, Newton described an experiment in which water effluxing horizontally from a hole 20 inches off the ground, strikes the ground 37 inches away, compared with a theoretical distance of 40 inches for a parabolic trajectory in the absence of air resistance, assuming the new value for the efflux velocity.[28]

The issue of the efflux velocity did not end so simply, however. When Cotes received Newton's proposed replacement for the efflux solution in 1710, he challenged it, calling attention to an experiment by Mariotte.[29] The efflux velocity Mariotte had inferred from the volume of water flowing out of a given hole in a given time was the same as Newton's original value. The need to reconcile these conflicting experiments led Newton to perform a further experiment in which he found that the diameter of an effluent stream $\frac{1}{2}$ an inch below a $\frac{5}{8}$-inch diameter hole is $\frac{21}{40}$, not $\frac{25}{40}$, of an inch. From this, he concluded that the area of the stream is $\frac{1}{\sqrt{2}}$ of the area of the hole, so that the true efflux velocity in Mariotte's experiment is $\sqrt{2}$ times greater than Mariotte had thought.[30] In effect, Newton had inferred an efflux coefficient of 0.71 for a circular orifice; the modern value is 0.62, implying an error in Newton's reported measurement of only 0.033 inches, just a little more than $\frac{1}{40}$ of an inch.

The factor of $\sqrt{2}$ increase in the efflux velocity implied a factor of two reduction in Newton's original theoretical value for the dimensionless inertial resistance acting on the front of a sphere. The new value for the resistance on the front face thus corresponded to a drag coefficient of 1.0 instead of 2.0.

4.2 *The vertical-fall experiments*

Precisely when Newton first performed vertical-fall experiments in water is not altogether clear, but it was surely before he moved to London.[31] An anecdote relayed by Conduit tells of the glass window at the bottom of a $9\frac{1}{2}$-foot trough breaking during an experiment, inundating Newton.[32] In one of his 1694 memoranda in the *Correspondence*, Gregory says that Newton "considers less certain the conclusions he has reached about the resistance of liquids from experiments with the pendulum, but he will arrange analogous experiments for falling bodies (but made of paper so that the resistance may become perceptible)."[33] A memorandum from two months later adds more details:

> "He will make many changes in Sec. 7, Book II, particularly in Proposition 37 about the height of jets of water where he adjusts his errors and proves everything more easily: also in the General Scholium appended to Proposition 40. For he does not sufficiently trust the observations obtained from a pendulum to determine the ratio of the resistance, but he will subject everything afresh to the test of falling weights. He is choosing the place for contriving his experiments from the top of Trinity College Chapel into his own garden on the right as one enters the College."[34]

Newton's vertical-fall experiments in air were ultimately carried out by Francis Hauksbee in the newly completed St. Paul's Church in London in 1710.[35] A Gregory memorandum gives us reason to think that most of the vertical-fall results in water reported in the second edition of the *Principia* were obtained after 1707, while Newton was actively engaged in this edition.[36]

Only three of the vertical-fall results in water reported in the second edition were from the $9\frac{1}{2}$-foot trough. Perhaps the failure of the glass curtailed this round of experiments. Newton rejected the data from one of these three, in which the very low buoyant weight of the falling spheres made the results excessively sensitive to small errors in their weight. The other two experiments showed significantly lower levels of resistance than the pendulum-decay experiments had shown; in modern terms, they give drag coefficients of 0.51 and 0.52.[37] The later nine experiments in a 15-foot deep trough of water that were reported in the second edition gave results generally very close to these. Hauksbee's results for a 220-foot vertical-fall in air gave virtually the same dimensionless magnitude for air.

The conclusion was clear: the resistance forces inferred from the pendulum-decay experiments were 35% and more too high. Newton attributed this, probably correctly,[38] to the pendulum's inducing a to-and-fro motion of the fluid around it, increasing the relative velocity between the fluid and the bob.

In modern terms, then, the correction to the efflux problem and the shift to relying on vertical-fall experiments left Newton with a measured drag coefficient for spheres around 0.5 and a theoretical inertial resistance on their front face of 1.0. If he had followed the logic of the first edition, he would have concluded that the counteracting force induced by the action of the fluid on the rear is around half the force on the front. Instead, he abandoned this logic, revising his theoretical value for the force on the front face to what amounts to a drag coefficient of 0.5, in near agreement with the experimental values in vertical-fall. Why did he do this?

5 Resistance Forces in the Second (and Third) Editions

Newton moved the General Scholium on the pendulum-decay experiments from the end of Sec. 7 to the end of Sec. 6 in the second edition, cutting it in the process. There, it serves the modest purpose of establishing that the v^2 resistance term dominates all others and varies, at least to a reasonable approximation, as $\rho_f d^2$. The comparison between the experimentally determined magnitude of the inertial resistance from pendulum-decay and the values obtained from theoretical models was eliminated, and with it the conclusion about the counteracting effect of the action of the fluid on the rear face.

The first half of Sec. 7 was modified a little, and the second half was replaced. Proposition 34, which announced that the similarity results for rarified fluids in Propositions 32 and 33 hold for highly lubricitous continuous fluids as well, was eliminated, most likely to allow Newton to retain the original proposition numbers in the remainder of Book II. The theoretical analysis of the inertial force in rarified fluids then became Proposition 35; the magnitude of this force remained the same as in the first edition, though the text was extensively revised, dropping the original non-dimensionalization in favor of the one employed for continuous fluids. The rest of Sec. 7 was rewritten entirely, starting with the new solution of the efflux problem, now Proposition 36. At the end of the section, Newton appended to the new Proposition 40, a Scholium presenting the results of 12 vertical-fall experiments in water and one in air; a second in air, carried out by Desaguliers in 1719 (again in St. Paul's), was added in the third edition.

5.1 *The new theory for continuous fluids*

The new solution for the efflux problem employs the notion of a cataract of flow, as displayed in Fig. 2. The funneling of the flow in the cataract gives Newton a way of reasoning to the experimentally known efflux velocity, and the oblique angle of the flow entering the hole at its edge lets him introduce the measured contraction of the effluent stream. Newton's reasoning here, employing a thought-experiment stretching over several pages, has been subjected to a good deal of ridicule, more I think than it deserves.[39] Admittedly, the reasoning is *ad hoc* and somewhat contrived; the explanation of the velocity is physically unilluminating, especially when compared with Daniel Bernoulli's solution a couple of decades later;[40] and the true cause of the contraction of the effluent stream is a viscously induced vortex around the orifice. Still,

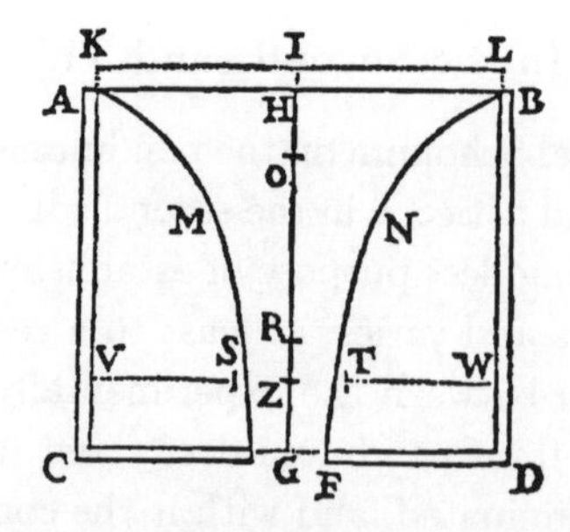

Fig. 2 The efflux problem diagram from the second edition of Newton's *Principia*. The curved lines BNF and AMS represent streamlines producing a constriction of the efflux stream.

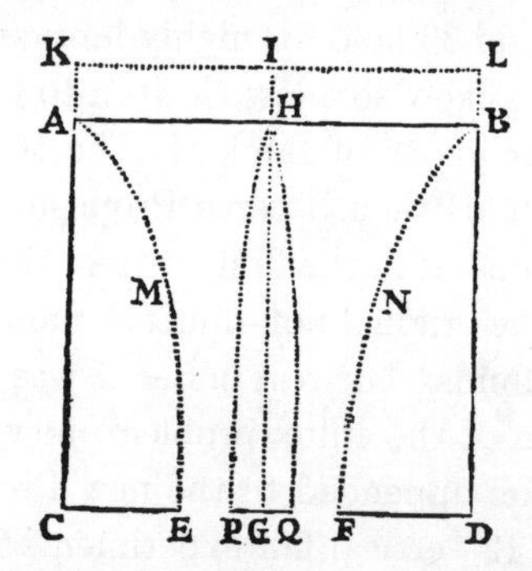

Fig. 3 Newton's diagram for the column of "continuous" fluid borne by the moving body PGQ. The lines HQ and HP defining the inner flow boundary are curved for the same reason as are lines BNF and AME defining the outer flow boundary.

much of Newton's discussion is devoted to complications arising from the two-dimensionality of the flow, complications that have required lots of attention in modern wind-tunnel design. Newton's new solution deserves a paper unto itself.

In a sequence of nine corollaries to the efflux solution, Newton developed a new result for the weight of the column of fluid bearing on a disk in the middle of an effluent stream. In the same way that the outer boundary of the cataract, BNF in Fig. 3, must be convex, the inner boundary defining the column of fluid acting on PQ must be convex; hence the weight of this column must be greater than that of a cone of fluid above PQ, i.e. greater than $\frac{1}{3}$ of the weight of the cylinder above PQ. Similarly, the angle of the flow at Q, just as that at F, must be acute; hence the weight of the column must be less than that of a hemispheroid with surface HQ perpendicular to PQ, i.e. the weight must be less than $\frac{2}{3}$ of the weight of the cylinder above PQ. Newton then concluded that the weight of the column is "very nearly equal to" $\frac{1}{2}$ the weight of the cylinder above PQ, "for this weight is an arithmetical mean between the weights of the cone and the hemispheroid." This weight is $\frac{1}{2}$ the value he had proposed in the first edition. The new efflux velocity drops the dimensionless inertial resistance on the disk by a factor of two. The new weight of the column above the disk drops it by an additional factor of two. This is how Newton's theoretical value for the inertial resistance on the front face dropped by a factor of four from the first to the second edition.

The question of the effects of the difference in shape between a disk and a sphere and the question of the action of the fluid on the rear face of the body still remain. Newton addressed these questions in the corollaries to Propositions 37 through 39 and in four lemmas and three Scholia accompanying these propositions. There, he invoked two-dimensional features of the flow and a postulated instantaneous propagation of pressure in continuous fluids to argue that both of these effects are at most second order. The upshot is that the theoretical value of the net inertial resistance on a sphere moving in a continuous fluid — or for that matter, on a cylinder moving lengthwise or a disk[41] — amounts to a drag coefficient of 0.5, "very nearly."

Newton's reasoning in relegating the effects of shape and the action of the fluid on the rear to second order continues and supplements the thought-experiment in the new efflux solution. Even so, this reasoning is different from the reasoning to his new conclusion about the weight of the fluid on a body in the efflux stream. He invoked experimental results on the height reached by the efflux stream and the width of the stream in the orifice to support his conclusion about the weight of the fluid. Thus, even if his theoretical reasoning in reaching this conclusion was wrong, readers at the time had reasons to trust the conclusion. By contrast, he offered no independent experimental evidence to support his conclusions about the effects of shape and the action of the fluid on the rear. As a consequence, his argument for these conclusions invited reservations, underscoring the advantages of a unified solution that treats the action of the fluid over the entire body in a single sweep.[42]

5.2 *The new comparison between theory and experiment*

As Newton's new theoretical value for the inertial resistance on spheres was prompted in part by his initial vertical-fall experiments in water, he was not

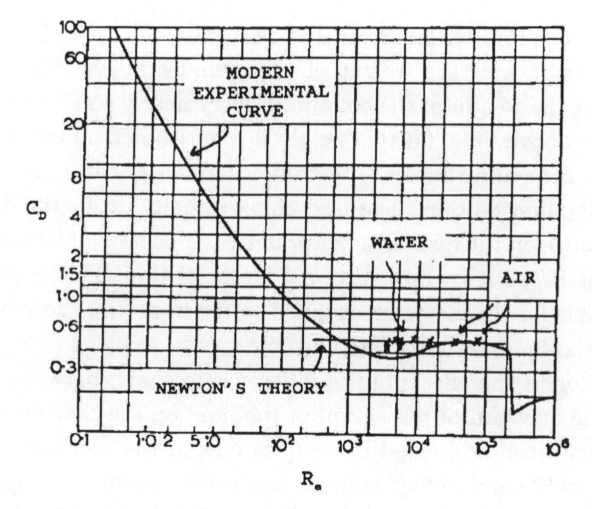

Fig. 4 Drag coefficient versus Reynolds number for spheres — a modern assessment of Newton's theory of resistance and vertical-fall data from the 2nd and 3rd editions of the *Principia.* The modern experimental curve is derived from wind-tunnel and water-table measurements of the resistance forces on spheres at velocities for which compressibility effects are negligible.

surprised to find it in good agreement with the results of his further vertical-fall experiments. Figure 4, which plots his experimental results against our modern Reynolds number,[43] displays this agreement.[44] The figure also shows that the experimental results for the resistance forces on spheres reported in the *Principia* are in impressively good agreement with modern measured values — a fact noted by R.G. Lunnon here at the Royal Society some 70 years ago.[45] The vertical-fall experiments, especially those carried out by Hauksbee and Desaguliers in St. Paul's, were of high quality by any standards.

Even if we grant Newton that the weight of the fluid acting on a disk in an efflux stream is between $\frac{1}{3}$ and $\frac{2}{3}$ of that of the cylinder above it, his choice of $\frac{1}{2}$ is physically arbitrary. Given the results of his vertical-fall experiments, this choice forced Newton to relegate the action of the fluid on the rear face to second order. Any other choice between $\frac{1}{3}$ and $\frac{2}{3}$ would have allowed the experiments to answer questions about the action on the rear, just as the experiments were allowed to in the first edition. Why, then, did Newton insist on precisely $\frac{1}{2}$?

I do not know the answer to this question. The best I can do is to list some possibilities. One possibility is that Newton wanted to tighten his argument against Descartes, i.e. the argument from the absence of resistance encountered by comets to the conclusion that interplanetary space is empty. As we saw above, Newton recognized that the theoretical possibility of a counteracting fluid force on the rear of moving bodies gave Cartesians a way of maneuvering around this argument. Many readers have thought that the main point of Book II was to argue against Descartes.

Another possibility is that Newton felt more need to secure close agreement between theory and experiment in the second edition than he had in the first. This is not the only place in which the second edition abandons a solution that had been put forward as a first approximation in the earlier edition, with the thought that measurements could then answer questions that would lead to a more refined approximation. Much the same happened with the solution for the figure of the earth and the variation of gravity with latitude in Propositions 19 and 20 of Book III.[46]

A third possibility is that Newton himself was persuaded by the reasoning given in the *Principia* that the force induced on the rear of moving bodies is, normally, insignificant. His choice of $\frac{1}{2}$ could then have been inferred from the vertical-fall data without begging any questions. The issue this possibility raises is whether the reasoning given in the text was something more than a mere contrivance. This reasoning proceeds from the assumption that pressure disturbances in a continuous fluid propagate "infinitely swifter" than the moving body, so that the fluid pressure on the rear face remains the same whether the body is moving or not. Such reasoning ignores recirculating flows of the sort that are glaringly evident behind, for example, bridge abutments. Regardless of whether Newton was persuaded by his reasoning, those who followed him, as we shall see, were not.

A fourth possibility — the one I am inclined to take most seriously — stems from a remark Newton made in the transition from the theoretical result in Proposition 40 to the vertical-fall experiments:

> "This is the resistance that arises from the inertia of matter of the fluid. And that which arises from the elasticity, tenacity, and friction of its parts can be investigated as follows."[47]

Given the failure of the pendulum-decay experiments to disaggregate the different effects contributing to resistance, the second edition is offering a new way of doing this. If the dominant inertial effect can be calculated precisely beforehand, and the sphere "encounters another resistance in addition, the descent will be slower, and the quantity of this resistance can be found from the retardation."[48]

Viewed in this light, Newton's choice of $\frac{1}{2}$ is best regarded as a *working hypothesis*, a first approximation on which further research can be predicated. The choice of $\frac{1}{2}$ need not then have been mere wishful thinking. Evidence would have accrued it to the extent that the further research predicated on it would have succeeded in yielding stable results on the magnitude and variation of non-inertial mechanisms of resistance. Failure of such research to yield stable results would have given grounds for modifying or abandoning the $\frac{1}{2}$ number. This strategy might well have seemed the best hope not only for investigating the non-inertial mechanisms, but also for providing a basis for refining the $\frac{1}{2}$ number. Admittedly, this is attributing a very sophisticated research strategy to Newton. As I have argued elsewhere,[49] however, it does not seem to me all that different from research strategies he adopted at other points in the *Principia*, for example with the problem of the Moon.

One final point needs to be made in regard to this last possibility. As Fig. 4 indicates, the forces inferred from two of the vertical-fall experiments fall *below* Newton's theoretical value. Furthermore, Newton reported and then discarded the data from the three experiments in which the resistance forces were most in excess of his theory. These happened to be the ones in which the velocities were greatest. Newton attributed this excess in part to the action of the fluid on the rear of the sphere: "the swifter the balls, the less they are pressed by the fluid in back of them; and if the velocity is continually increased, they will at length leave an empty space behind, unless the compression of the fluid is simultaneously increased."[50] The data accordingly raised the worry that the action on the rear, even if second order, might be comparable in magnitude not just to the inaccuracies in the measurements, which in principle could be improved, but to the non-inertial effects themselves. If so, the proposed research strategy was going to require a new theoretical solution in which the action on the rear, second order or not, would be taken into account.

6 In the Wake of the *Principia*

The same number of years elapsed between the first and second editions of the *Principia* as between the third edition and d'Alembert's *Essay on a New Theory of the Resistance of Fluids*.[51] d'Alembert opens his essay with a critical review of Newton. He says that Newton's theory of inertial resistance in rarified fluids is "a research of pure curiosity,... not applicable to nature."[52]

Of Newton's solution for inertial resistance in continuous fluids, he objects that it is "intended to elude rather than surmount the difficulty of the problem."[53] The difficulty of the problem, needless to say, is determining the action of the fluid everywhere on the surface of the moving body.

d'Alembert made sure that his solution did not elude this difficulty. He devised partial differential equations for inviscid, incompressible fluids and solved them for the pressure distribution on the surface of a sphere. When he integrated this pressure over the entire surface, he discovered to his great surprise that the net inertial resistance force is exactly zero! This raised the question of other shapes, for which he found the same result: the resistance force on a body moving in a truly inviscid, incompressible fluid is invariably exactly zero, regardless of shape. In other words, the correct result to the problem Newton set himself in the last part of Sec. 7 is zero.[54] The fluid force induced on the rear of a body in what we now call an *ideal fluid* is always exactly equal and opposite to the fluid force on the front. This fundamental result in fluid mechanics is known as d'Alembert's paradox. The paradox was not resolved until the first decade of the twentieth century.[55]

I often wonder how Newton would have reacted to d'Alembert's paradox. He surely would have been far from pleased. For one thing, though it was ceasing to matter by 1752, d'Alembert's result undercut his argument against Descartes: if the Cartesian celestial fluid is inviscid, then the absence of resistance on comets shows nothing. d'Alembert's result would also have undercut the research strategy proposed above: discrepancies between measurements and the theoretical value for pure inertial resistance were never going to provide any basis at all for investigating non-inertial contributions to resistance. The interesting question, however, is whether Newton would have taken d'Alembert's result to be showing — rightly, I might add — that Newton's entire approach to resistance forces was fundamentally misconceived. However appropriate such a model may be for giving rough practical values, resistance forces simply cannot be represented as the sum of a few independent terms in powers of velocity, one term for purely inertial effects, another for purely viscous ones, etc., when the goal includes getting at the physics underlying the forces.[56] Resistance forces result from intertwined inertial and viscous actions. Giving a full theoretical account of this intertwining remains as one of the unsolved problems of physics.[57]

Would Newton have seen d'Alembert's result as showing that the investigation of resistance forces was going to have to be re-started from square one?

This is a question that the correspondence and manuscripts, voluminous as they are, are unlikely to answer.

7 Afterword: A Note of Appreciation

Neither Rupert Hall nor Tom Whiteside have written on Sec. 7 of Book II of the *Principia.* Their efforts on Book II have focused on the mathematical solutions for motion in resisting media in Secs. 1–6, not on Newton's attempts to combine these solutions with experimental results to reach conclusions about the forces of resistance. As the notes to this paper has made clear, however, it would have been difficult to manage without their writings. As a philosopher, I turned to Newton's *Principia* hoping to learn how we first came to have high quality evidence in science. I would have abandoned the project early on were it not for Tom Whiteside's notes in Volume 6 of the *Mathematical Papers.* Book II of the *Principia* is important to me because it shows Newton trying to marshall evidence in an area of physics that has proved to be notoriously intractable. The first book I read on seventeenth century treatments of fluid resistance was the published version of Rupert Hall's doctoral dissertation, *Ballistics in the Seventeenth Century.* Sorting out what went on between the first and second editions of Book II would have been hopeless without the *Correspondence.* Rupert and Tom, and their American counterparts, Bernard Cohen, Curtis Wilson and the late Sam Westfall, have given people like me a degree of access to Newton's science that we would never have had without them. The occasion of this paper allows me to acknowledge this and to express, for myself and many others like me, our gratitude.

Notes and References

1. Newton used two different non-dimensionalizations for the resistance force in the first edition, one for rarified fluids and the other, stated ambiguously, for continuous fluids. In the second edition, he used a disambiguated form of the latter for both types of fluid: in a rarified fluid "a sphere encounters a resistance that is to the force by which its whole motion could be either destroyed or generated, in the time in which it describes $\frac{2}{3}$ of its diameter by moving uniformly forward, as the density of the medium is to the density of the sphere." There is nothing, as such, wrong with this way of normalizing the force. Its only notable shortcoming is that it has to include the density of the moving body as a factor, while the modern method does

not. I have elected to replace it with the modern method partly as a matter of convenience, but mostly to facilitate comparisons with modern data.

2. The drag coefficient varies inversely with Newton's number of diameters the sphere moves in the time its motion can be generated or destroyed. A drag coefficient of 1.0 for a sphere corresponds to a distance for Newton of $\frac{4}{3}$ of its diameter, and a drag coefficient of 0.5 corresponds to $\frac{8}{3}$.
3. Newton expressed the view that resistance forces most likely involve these three terms in the Scholium ending Sec. 3 of Book II. There he attributed the first term to the "tenacity" of the fluid, the second, to its "friction," and the third to its "density." The first term corresponds to simple surface friction between the fluid and the moving body; this term has been dropped from modern treatments of fluid resistance, in keeping with the "no-slip" condition at the fluid-surface interface introduced by Stokes. The second term, which Newton at times spoke of as representing the "attrition of the parts" of the fluid, corresponds to the viscous term, and the third term, to the inertial term, in modern engineering models of fluid resistance.
4. In the final propositions of the tract *De Motu Corporum*, Newton proposed that the deceleration caused by resistance proportional to v varies with the density of the medium and the surface area of the sphere, and inversely with its weight. ["*De motu corporum in gyrum*," in *The Mathematical Works of Isaac Newton* (ed.) D.T. Whiteside (Cambridge, Cambridge University Press, 1974), Vol. 6, pp. 30–75.] In the *Principia*, Newton dropped this proposal and never said anything at all about how the v term might vary. In Query 28 of the *Opticks*, however, he remarked in passing that the resistance force on a sphere arising from "the attrition of the parts of the medium is very nearly as the diameter, or, at the most, as the *factum* of the diameter, and the velocity of the spherical body together." [*Opticks* (New York, Dover, 1952), p. 365.] He offered no explanation for this claim. In the query, it serves only as a step in the argument that the resistance on bodies "of a competent magnitude" arises almost entirely from the inertia of the fluid and not from the attrition of its parts; this argument in turn is a step in the larger argument of the query against the existence of an aetherial medium. Stokes's law for the purely viscous resistance on a sphere equates the force to $3\pi\mu dv$, where μ is the viscosity of the fluid.
5. Newton's celebrated alternative to a solution for projectile motion under resistance varying as v^2, Proposition 10 of Book II, is discussed in D.T.

Whiteside, "The mathematical principles underlying Newton's *Principia Mathematica*," *Journal for the History of Astronomy* **1** (1970): 126–131. A.R. Hall, discusses Newton's solutions for vertical-fall and projectile motions extensively in *Ballistics in the Seventeenth Century* (Cambridge, Cambridge University Press, 1952), Chap. 6.

6. *Mathematical Papers* **6** (17): 448.
7. Newton distinguished a third type as well, *elastic* fluids, but he did not attempt to give a theoretical magnitude for the inertial resistance for them. His conception of an elastic fluid is that it is like a rarified fluid, but with forces among the particles. As a result of these forces, he cannot treat the particles as acting on a body independently of one another, as in a rarified fluid. Hence, for purposes of determining a theoretical value for inertial resistance, elastic fluids are more like continuous ones. Newton generally regarded air as an elastic fluid. As will be discussed below, however, when he compared experiments in air with theory, he ended up using the theoretical value of inertial resistance for continuous fluids.
8. *Isaac Newton's Philosophiae Naturalis Principia Mathematica* (eds.) Alexandre Koyré and I. Bernard Cohen, Vol. 2, p. 775f. Specifically, Newton gave a geometric characterization of the fraction of the initial motion lost while describing a given space.
9. *Ibid.*, p. 776.
10. See Whiteside, *op. cit.*, pp. 456–480.
11. In Proposition 37 in the second edition, however, Newton contrasted continuous with elastic fluids when arguing that the force arising from the action of the fluid on the rear of a moving body is negligible.
12. Newton qualified this conclusion in a further remark: "All these things are to be understood for a most subtle fluid. For if the water consists of thicker parts, it will flow out more slowly..., especially if the hole through which it flows is narrow." (Koyré and Cohen, *op. cit.*, p. 778).
13. *Ibid.*, p. 782. The magnitude of the inertial force acting on the front surface of the sphere given in the first edition is definitely greater than the magnitude given in the second edition. How much greater is a matter of some confusion. The magnitude given in the second edition is equivalent to a C_D of 0.5. The text of Proposition 38 in the first edition concludes,

   > "And therefore the resistance on a globe progressing uniformly in any very fluid medium whatever, in the time in which the globe describes $\frac{2}{3}$ of its own diameter, is equal to the force that, uniformly impressed

> upon a body of the same size as the globe and the same density as the medium, in the time in which the globe by progressing described $\frac{2}{3}$ of its own diameter, could generate the globe's velocity in that body."

If we take the phrase, "by progressing described $\frac{2}{3}$ of its own diameter," at the end of this statement to mean *progressing uniformly at its acquired velocity*, then the time is $(\frac{2}{3})d \div v = (\frac{2}{3})(d/v)$ and the magnitude given for the resistance corresponds to a C_D of 2.0. If, instead, we take this phrase to mean, *progressing from rest to its acquired velocity*, then the time is $(\frac{2}{3})d \div (v/2) = (\frac{4}{3})(d/v)$ and the magnitude corresponds to a C_D of 1.0. The first corollary to the proposition says that this resistance is the same as that obtained for a rarified fluid in which the particles are inelastic and hence do not rebound; this implies that the correct reading is the one giving a C_D of 1.0. The comparison with the pendulum experiments, however, says that the measured inertial component is around $\frac{1}{3}$ of the magnitude obtained in the proposition; this implies a C_D of 2.0. Complicating the matter further are a number of errata Newton listed for these two passages in one of his copies of the first edition (*ibid.*, p. 782 and p. 462); these errata seem to imply a C_D of 1.0. The phrasing Newton adopted in non-dimensionalizing the inertial resistance force in the second edition removes the ambiguity.

14. *Ibid.*
15. *Ibid.*
16. *Ibid.*
17. For the derivation of this expression in the manner of Newton and an analysis of its relationship to Proposition 30, see Whiteside *op. cit.*, p. 448, n. 17. A modern derivation can be found in Michael Nauenberg's appendix to the present paper. In the *Principia*, Newton approximated $\frac{2}{\pi}$ by $\frac{7}{11}$.
18. Newton used at least two ways in assessing his data: (1) he calculated the residual errors for the data points not used in the algebraic determination of the coefficients A_i; and (2) he calculated consecutive ratios of the arc lost, which should monotonically approach 4 if the term with the highest exponent is v^2. It is unclear whether he calculated values for the A_i using different combinations of data points. Of course, at that time no statistical methods were available for using all the data points in determining values for the coefficients.

19. The four combinations that we know Newton tried were $A_0 + A_1V + A_2V^2$; $A_1V + A_2V^2$; $A_{\frac{1}{2}}V^{\frac{1}{2}} + A_1V + A_2V^2$; and $A_1V + A_{\frac{3}{2}}V^{\frac{3}{2}} + A_2V^2$. A reason for thinking that he started with the first of these is that it is the only one for which he expressly derived values of the coefficients in Eq. (5) in the text of the *Principia*, specifically in the corollary to Proposition 30.
20. Newton used the second, fourth and sixth data points in the first data set to determine the values of A_i listed in the *Principia.*
21. Newton indicated in the *Principia* that one of his two experiments comparing resistance forces in water and mercury yielded a ratio of about 13 or 14 to 1, in good agreement with the density claim. He provided no data from this experiment, however, and he said that the second experiment indicated a smaller ratio, attributing this to a trough that was too narrow in comparison to the diameter of the bob. While the experiments comparing air and water yielded less than ideal results, the refinements that Newton introduced to improve those in water, and that he then included in his experiments in mercury, display how ingenious he was in designing and developing experiments.
22. My "Newton's experiments in fluid resistance" (in preparation) will provide a more detailed discussion of both his pendulum-decay and his vertical-fall experiments.
23. See Koyré and Cohen, *op. cit.*, p. 460f for the two paragraphs in which he compared theory with the pendulum experiments and drew conclusions about the action of the fluid on the rear of the moving sphere. These paragraphs were dropped in the second edition.
24. Specifically, Newton used the result from the case in which the pendulum lost $\frac{1}{8}$ of its 64-inch initial arc to conclude that "a globe of water moving in air encounters resistance that would cause it to lose $\frac{1}{3261}$ part of its motion during the time it describes the length of its semi-diameter." (The fraction was changed to $\frac{1}{3342}$ in the second edition.) Using the modern ratio of the density of water to air, 831, this is equivalent to a drag coefficient of 0.68. Newton expressed uncertainty in the first edition about this ratio of densities, allowing a value as low as 800 and as high as 850; the drag coefficients corresponding to these two values are 0.654 and 0.695, respectively.

    Two points are worth noting about the specific case Newton used in comparing theory and experiment. First, the disaggregated v^2 component in this case is 92% of the total resistance. Second, the consecutive ratios displayed in Table 2 indicate that the value he obtained for the v^2 component would have been greater had he used the fifth rather than the sixth data point.

25. Koyré and Cohen, *op. cit.*, Vol. 1, p. 460.
26. *The Correspondence of Isaac Newton* (ed.) H.W. Turnbull (Cambridge, Cambridge University Press, 1961), Vol. 3, p. 169.
27. Letter from Newton to Cotes, dated 30 September 1710; *The Correspondence of Isaac Newton* (eds.) A. Rupert Hall and Laura Tilling (Cambridge, Cambridge University Press, 1975), Vol. 5, p. 70.
28. *Principia*, Book II, Proposition 36 (case 3). Descartes, by the way, had solved this problem of the parabolic trajectory of an efflux stream in a letter to Constantin Huygens of 18 or 19 February 1643 [see Charles Adam and Paul Tannery (eds.), *Oeuvres de Descartes*, 2nd Ed. (Paris, J. Vrin, 1964–1974), Vol. 3, p. 621].
29. Letter from Cotes to Newton dated 21 September 1710, in *Correspondence*, Vol. 5, pp. 65–67. Newton's initially proposed replacement for the efflux solution has apparently been lost.
30. See *Principia*, Book II, Proposition 36.
31. R.S. Westfall concluded that the initial experiments in water were done while the first edition was in progress (Richard S. Westfall, *Never At Rest: A Biography of Isaac Newton* (Cambridge, Cambridge University Press, 1980), p. 455). If so, it seems curious that Newton did not take the results to be showing that the resistance forces inferred from the pendulum-decay experiments were excessive.
32. *Keynes MS*, 130.15, quoted from Westfall, *op. cit.*, p. 455.
33. Memorandum by David Gregory, dated 4 May 1694, in *Correspondence*, Vol. 3, p. 317f. The memorandum continues, "Also from projectiles from a catapult.... He believes that he now knows the curve in which a heavy body tending to the centre of forces according to the inverse square of the distance, and to which the resistance is as the square of the velocity, moves when obliquely projected." I know of no Newton manuscript proposing a solution to this problem.
34. Memorandum by David Gregory, dated c. July 1694, *Ibid.*, p. 384.
35. Journal Book (Copy) of the Royal Society, 10, 7 and 14 June 1710, p. 243. Hauksbee, who had been elected a Fellow in 1705, had become the principal demonstrator of experiments in the Royal Society following Hooke's death in 1703; it was thus appropriate that he carried out the experiments in

St. Paul's for Newton. Vertical-fall experiments in air from any lesser height than St. Paul's, by the way, would have had little chance of yielding good data, given the then existing limits on the accuracy of determining the time of fall using $\frac{1}{4}$ sec pendulums.

36. A memorandum of David Gregory, dated 15 April 1707, remarks, "The only two things that Sir Isaac Newton further desires, to make a new edition of Princ. Math. Philos. Nat. are these; First about the Resistance of Fluids in Lib. II about which he is affray'd he will need some new experiments. The second is concerning the Procession of the Æquinoxes Lib III of which he thinks he is fully master, but has not yet written it out. He does not think to go about these things, at least the Edition, for two years yet." [*David Gregory, Isaac Newton, and their Circle: Extracts from David Gregory's Memoranda* (ed.) W.G. Hiscock (Oxford, Oxford University Press, 1937), p. 40.]
37. The velocities in these two experiments were significantly different — 28 in $\sec^{-1}$ in the first experiment and 7.5 in $\sec^{-1}$ in the second — corresponding to Reynolds numbers of 17 440 and 4430, respectively. The result in the first is very close to the modern measured value, while the result in the second is on the high side (see Fig. 4).
38. The pendulum-decay experiments have a large number of likely sources of error. Based on Stokes's subsequent efforts, however, the principal source of error is almost certainly the to-and-fro motion of the surrounding fluid induced by the motion of the pendulum. See George Gabriel Stokes, "On the effect of the internal friction of fluids on the motion of pendulums," (1851), in *Mathematical and Physical Papers* (Cambridge, Cambridge University Press, 1901), Vol. 3, pp. 55–62.
39. The most prominent ridicule of the new efflux solution was by Clifford Truesdell in his "Reactions of late baroque mechanics to success, conjecture, error, and failure in Newton's *Principia*," in *Essays in the History of Mechanics* (New York, Springer-Verlag, 1968), pp. 138–183. Truesdell was not alone in this regard, however. For example, see Westfall, *op. cit.*, pp. 709–712.
40. Daniel Bernoulli obtained the correct efflux velocity from the conservation of *vis viva*, in the process formulating the basic principle of fluid mechanics now known as Bernoulli's formula:

$$\frac{p}{\rho} + \frac{1}{2}v^2 + \phi = \text{constant}$$

where ϕ is the potential per unit mass from external forces, and the constant can change from one streamline to the next.

41. de Borda's experimental refutation of Newton's theory turned on the claim that a sphere and a disk of the same frontal area encounter the same resistance in water. He used a rotating rig to compare the resistance on a sphere with that on a hemisphere with its flat face forward, finding (correctly) that the resistance on the latter is more than twice as great. Earlier he had shown that Newton's shape of least resistance in rarified fluids does not conform in any way with water or air. He concluded that "the ordinary theory of the impact of fluids only gives relationships which are absolutely false and, consequently, it would be useless and even dangerous to wish to apply this theory to the craft of the construction of ships." See Le Chevalier de Borda, "Expériences sur la résistance des fluides," *Mémoires de l'Académie Royale des Sciences*, pp. 358–376 (1763), and "Expériences sur la résistance des fluides," *ibid.*, pp. 495–503 (1767).
42. Remarks Newton himself made in passing in this reasoning also invited questions. For example, he called attention to the non-instantaneous propagation of pressure in elastic fluids like air. Yet he proceeded a few pages later to treat the action of the fluid at the rear of a sphere moving in air as negligible.
43. The Reynolds number is a dimensionless measure of the ratio of inertial to viscous effects for fluids. In the case of fluids interacting with spheres it is equal to $\rho_f dv/\mu$, where μ is the viscosity of the fluid. The Reynolds number is the generic similarity parameter for fluids in which compressibility effects and free-surface effects are negligible.
44. Newton in fact compared measured time and distance in vertical-fall with values he calculated, using Proposition 9 of Book II (vertical-fall with resistance as v^2), extended to a tabular solution in Proposition 40. (A derivation of the tabular solution can be found in Michael Nauenberg's appendix to the present paper.) This tabular solution, and hence his calculated values, presupposed his theoretical value for resistance in a continuous fluid — in effect, a drag coefficient of 0.5 — attributed to *both* air and water. Table 3 summarizes his comparisons for the experiments in water. The drag coefficients shown in Fig. 4 have been inferred from the measured time and distances Newton lists, as indicated in the table. Specifically, the value of the drag coefficient in the modern solution for vertical-fall with resistance as v^2 was iteratively varied until the distances

and times of fall in this solution matched those given by Newton. These drag coefficients are accordingly independent of Newton's theory. As noted in the table, Newton introduced corrections to some of the raw data in order to compensate for the fact that his troughs were not, in effect, infinitely wide compared to the diameters of the spheres. As the table shows, the corrections make the agreement between theory and experiment, which is good to begin with, even better.

Table 3 Newton's data for the vertical-fall experiments in water.

Experiment	Diameter (in)	Weight (grns)	Buoyant Weight (grns)	Actual Time (sec)	Theoretical Fall (in)	Actual Fall (in)
1	0.892	156.3	77.0	4.0	113.06 112.08*	112.0
2	0.813	76.4	5.1	15.0	114.07 113.17*	112.0
4	0.999	139.3	7.1	25.0 (23.5–26.5)	184.25 181.86*	182.0

Experiment	Diameter (in)	Weight (grns)	Buoyant Weight (grns)	Actual Fall (in)	Theoretical Time (sec)	Actual Time (sec)
5	1.00	154.5	21.5	182.0	14.5	14.8–16.5
8	0.999	139.2	6.5	182.0	26.0	25.0–26.0
11	0.693	48.1	3.9	182.5	23.3	21.8–23.0
12	1.010	141.1	4.4	182.0	32.3	30.5–32.5
6	1.00	212.6	79.5	182.0	7.5	7.5–9.0
9	0.999	273.5	140.8	182.0	5.7	6.0–6.5
10	1.258	384.4	119.5	181.5	7.8	8.9–9.5
7	1.247	293.7	35.9	181.5	14.0	14.8–16.5
3	0.967	121.0	1.0	112.0	40.0	46, 47, 50

*After correction based on Proposition 39.

45. R.G. Lunnon, "Fluid Resistance to Moving Spheres," *Proceedings of the Royal Society of London*, Series A, Vol. 10, pp. 302–326 (1926).
46. In the first edition, Proposition 20 of Book III expressly proposed that deviations from the calculated values of the variation in surface gravity

with latitude be used to infer how the density of the earth varies radially from the surface to the center. This proposal disappeared in the second and third editions.

47. *Principia*, Book II, Proposition 40.
48. *Ibid.*
49. "The Newtonian style in Book II of the *Principia*," *Isaac Newton's Natural Philosophy* (eds.) Jed Z. Buchwald and I. Bernard Cohen (Cambridge, MIT Press, 2000).
50. The quotation is from the summary discussion of the experiments in water following Experiment 12 in the Scholium appended to Proposition 40 of Book II.
51. Jean d'Alembert, *Essai d'une nouvelle théorie de la résistance des fluides* (Paris, David, 1752).
52. *Ibid.*, p. xiv.
53. *Ibid.*, p. xx.
54. Newton, of course, never spoke of inviscid fluids. That this is what he was considering is nonetheless clear, as shown in the following remark: "Some difference can arise from a greater or lesser friction; but in these lemmas we are supposing that the bodies are very smooth, that the tenacity and friction are nil...." (Scholium to Lemma 7, Book II).
55. The paradox was resolved, at least for engineering purposes, by Ludwig Prandtl's boundary layer theory, first presented in 1904. For a lucid account of boundary layers and the resistance forces on bodies moving in fluid media, see L. Prandtl and O.G. Tietjens, *Applied Hydro- and Aerodynamics*, tr. J.P. Den Hartog (New York, Dover, 1934).
56. A two-term model in v and v^2, the first term representing viscous effects and the second term, inertial, is standard in most engineering applications, as well as for defining corrections to compensate for resistance forces in many physics experiments. As is evident from Fig. 4, such a model can approximate the drag coefficient versus Reynolds number curve for spheres to a reasonable extent up to the onset of fully developed turbulence at Reynolds numbers above 10^5. [It does not do so well for other shapes, which often include more inflection points than the experimental curve for spheres displays; for more details, see Hermann Schlichting's *Boundary-Layer Theory*, 7th Ed. (New York, McGraw-Hill, 1979).]
57. In particular, the problem of turbulence — a phenomenon reflected by the abrupt drop in the drag coefficient at very high Reynolds numbers,

as displayed in Fig. 4 — remains unsolved. I quote from Stig Lundqvist's Introduction to Kenneth Wilson's Nobel Prize Lecture:

> "The development in physics is on the whole characterized by a close interaction between experiment and theory.... This close interaction between theory and experiment keeps the frontiers of physics moving forward very rapidly.
> However, there have been a few important exceptions, where the experimental facts have been well known for a long time but where the fundamental theoretical understanding has been lacking and where the early theoretical models have been incomplete or even seriously in error. I mention here three classical examples from the physics of the twentieth century, namely superconductivity, critical phenomena and turbulence.... The third classical problem I mentioned, namely turbulence, has not yet been solved, and remains a challenge for the theoretical physicists." [*Nobel Lectures in Physics: 1981–1990* (Singapore, World Scientific Publishing Company, 1993), p. 95.]

(Wilson received the Nobel Prize in Physics in 1982 "for his theory for critical phenomena in connection with phase transitions.")

Addendum to G. Smith's "Fluid Resistance: Why did Newton Change his Mind?"

MICHAEL NAUENBERG

Department of Physics, University of California
Santa Cruz CA 95064, U.S.A.

1 Perturbation Theory for Damping of a Cycloid Pendulum (Footnote 7)

Let $F_{\text{res}} = m\beta v^n$ be the damping force acting on a cycloid pendulum of length l and mass m. Setting the velocity $v = dx/dt$, where x is the arc length and $dv/dt = (1/2)dv^2/dx$, the equation of motion for the damped harmonic oscillator is given by

$$\frac{1}{2}\frac{dv^2}{dx} + \omega^2 x = -\beta v^n \tag{1}$$

where $\omega^2 = g/l$. Let

$$v^2 = \omega^2(x_0^2 - x^2) - \delta v^2 \tag{2}$$

where x_0 is the initial amplitude of the oscillation, and $v^2 = 0$, $\delta v^2 = 0$ at $x = x_0$. Then, to lowest order perturbation in powers of β,

$$\frac{d\delta v^2}{dx} = 2\beta\omega^n(x_0^2 - x^2)^{n/2}\,. \tag{3}$$

Hence, the change in δv^2 during half an oscillation (from x_0 to $-x_0$) is given by

$$\delta v^2_{\text{max}} = 4\beta\omega^n x_0^{n+1}\int_0^1 dx(1 - x^2)^{n/2}\,. \tag{4}$$

Setting $x = -x_0 + \delta x_0$ in Eq. (2), where δx_0 is the loss of amplitude in one half-oscillation due to damping (by requiring that $v^2 = 0$ at this point) gives

$$\delta x_0 = \frac{\delta v^2_{\text{max}}}{2x_0\omega^2} \tag{5}$$

or

$$\delta x_0 = 2\beta\omega^{n-2}x_0^n\int_0^1 dx(1 - x^2)^{n/2}\,. \tag{6}$$

Substituting $x_0 = v_{\max}/\omega$, we have

$$\delta x_0 = A_n v_{\max}^n \tag{7}$$

where

$$A_n = 2\frac{\beta}{\omega^2}\int_0^1 dx(1-x^2)^{n/2}. \tag{8}$$

Correspondingly, the resistance force can be written in the form

$$\frac{F_{\text{res}}}{mg} = \frac{\beta v^n}{\omega^2 l} = \frac{A_n v^n}{2l\int_0^1 dx(1-x^2)^{n/2}}. \tag{9}$$

[See also *The Mathematical Papers of Isaac Newton*, (ed.) D.T. Whiteside (Cambridge University Press), 1974, Vol. 6, p. 448.]

2 Fall of a Body in a Fluid with v^2 Resistance (Footnote 44)

For a body falling freely in a fluid, the gravitational acceleration g is reduced by a factor $(1-\rho_f/\rho)$ due to buoyancy, where ρ_f is the density of the fluid and ρ is the density of the body. If the resistance force $F_{\text{res}} = m\beta v^2$, then the equation of motion is

$$\frac{dv}{dt} = g' - \beta v^2 \tag{10}$$

where $g' = g(1-\rho_f/\rho)$. Hence,

$$\beta t = \int_0^v dv \frac{1}{(H^2 - v^2)} \tag{11}$$

where $H = \sqrt{g'/\beta}$ is the terminal velocity. Setting $G = 1/(\beta H)$, we have

$$t = \frac{G}{2}\ln\left(\frac{H+v}{H-v}\right) \tag{12}$$

and correspondingly

$$v = H\frac{(N-1)}{(N+1)} \tag{13}$$

where $N = \exp(2t/G)$.

Then, the height y of fall in time t is given by

$$y = \int_0^t dtv = H\int_0^t dt\left(1 - \frac{2}{(N+1)}\right) \tag{14}$$

or

$$y = Ht - F\left(2\ln 2 - 2\ln\frac{(N+1)}{N}\right) \tag{15}$$

where $F = HG/2$.

3 Newton's Solution

This solution [Eq. (15)] is given by Newton in Proposition 40 of the *Principia* in the form

$$y = \frac{2PF}{G} - 1.3862943611F + 4.605170186LF \tag{16}$$

where $P = t$ and $L = \log((N+1)/N)$. For this result, Newton referred the reader to Proposition 9, where he gave the relevant integrals as areas under hyperbolas, but did not carry out the relevant integrations. Moreover, the quantities P, G, H, F, L, N were not defined there, but were given in Proposition 40 as follows:

> "... D the diameter of the globe, F a space which is to $(4/3)D$ as the density of the globe is to the density of the medium ($F = (4/3)D(\rho/\rho_f)$), G the time in which the globe falling with the weight B without resistance describes the space F ($G = \sqrt{2F/g'}$), and H the velocity which the body gets by that fall ($H = g'G$). Then H will be the greatest velocity with which the globe can possibly descend with the weight B in the resisting medium.... Find the absolute number N agreeing to the logarithm (base 10) $0.4342944819(2P/G)$ (corresponding to $N = \exp(2P/G)$), and let L be the logarithm of the number $(N+1)/N$."

We now verify that Newton's definitions correspond to those given here. Newton assumed that $F_{\text{res}} = \rho_f A v^2$, where the area $A = \pi D^2/4$. Setting $F_{\text{res}} = (1/2)C_d\rho_f\pi D^2 v^2$, this implies that $C_d = 1/2$. Therefore, $\beta = F_{\text{res}}/mv^2 = (1/8)C_d\rho\pi D^2/m$, where $m = (\pi/6)\rho D^3$ is the mass of the "globe." Hence $\beta = (3/4)C_d(\rho_f/\rho)(1/D)$, and we have that $F = HG/2 = 1/(2\beta) = (2/3)(D/C_d)(\rho/\rho_f)$. For $C_d = 1/2$, this result corresponds to Newton's definition of F. (Note that C_d, which we now know depends on the Reynolds number $\rho v D/\mu$, contains the information about the viscosity μ which is hidden in Newton's expression, derived by him neglecting viscosity!) It follows that G, which we obtained in the form $G = 1/\sqrt{g'\beta}$, also gives Newton's relations $G = \sqrt{2F/g'}$ and $H = 2F/G = g'G$.

In the *Principia*, Newton wrote that

> "these things appear by Proposition IX, Book II, and its Corollaries, and are true upon this supposition."

However, his contemporaries would have found it difficult to obtain his result [Eq. (15)] by following the arguments of this proposition. It is interesting to discuss this proposition because it illustrates the strange ways, by modern standards, which Newton sometimes used to describe his calculus in the *Principia*, and hide some crucial results. In Proposition 9, Newton expressed the integral for the time of fall as a function of the velocity [Eq. (11)], in geometrical form as the area under a sector of a hyperbola, which in Cartesian coordinates has the form

$$y^2 - x^2 = H^2 . \tag{17}$$

Setting $v = Hx/y$, which corresponds in algebraic language to Newton's geometrical construction and in effect is a change of variables of integration, one readily finds that

$$\int_0^v \frac{dv}{H^2 - v^2} = \frac{1}{H}\int_0^x \frac{dx}{y} = \frac{2}{H^3}\left(\int_0^x ydx - \frac{1}{2}xy\right) . \tag{18}$$

The last form of the integral corresponds to the area under a sector of the hyperbola, as Newton proved geometrically in Proposition 9, but Newton did not even give a hint as to how to evaluate this integral. Indeed, it is more likely that he evaluated the integral in its original form [Eq. (10)], by the familiar procedure of factoring the integrand in the form

$$\frac{1}{(H^2 - v^2)} = \frac{1}{2H}\left(\frac{1}{(H+v)} + \frac{1}{(H-v)}\right) . \tag{19}$$

However, the original integral does not have a geometrical representation, and therefore could not be presented in the format of the *Principia*. It appears that Newton introduced the transformed integral mainly to conform to this format, and not because it provided him with a useful transformation to evaluate the required integral. In Proposition 40, he expressed his result with the comment

"... and the velocity falling will be $H(N-1)/(N+1)$..."

corresponding to Eq. (13).

In contrast, in Propositions 8 and 9, the integral for y as a function of v is expressed directly as an area under a hyperbola. In Newton's notation, this hyperbola is proportional to $1/(AC - AK)$, where $AK = v^2$ is the independent variable, and $AC = H^2$. Hence,

$$y \propto \int_0^{AK} \frac{1}{(AC - AK)} = \ln\left(\frac{AC}{(AC - AK)}\right) . \tag{20}$$

However, the constant of proportionality F, which determines the scale of y, is missing in Proposition 9. In our notation,

$$y = F \ln\left(\frac{H^2}{(H^2 - v^2)}\right) \tag{21}$$

and substituting for v Eq. (13) gives Newton's Eq. (15).

In Proposition 40 Newton gave a table of numerical values for t, v and y in units of G, H and F, respectively (Table 1). These values were presumably calculated by his power series expansion of the exponential and logarithmic functions. What is extraordinary in the table is the range of eight decimal places. To calculate $\exp(z)$ for large z, one needs approximately $n = z\exp(1.0)$ terms. Thus, n is about 54 for $z = 20$. I checked numerically that $n = 51$,

Table 1 Newton's computation in Proposition 40.

Tempora P	*Velocitates cadentis in fluido.*	*Spatia cadendo descripta in fluido.*
0,001 G	99999$\frac{29}{30}$	0,000001 F
0,01 G	999967	0,0001 F
0,1 G	9966799	0,0099834 F
0,2 G	19737532	0,0397361 F
0,3 G	29131261	0,0886815 F
0,4 G	37994896	0,1559070 F
0,5 G	46211716	0,2402290 F
0,6 G	53704957	0,3402706 F
0,7 G	60436778	0,4545405 F
0,8 G	66403677	0,5815071 F
0,9 G	71629787	0,7196609 F
1 G	76159416	0,8675617 F
2 G	96402758	2,6500055 F
3 G	99505475	4,6186570 F
4 G	99932930	6,6143765 F
5 G	99990920	8,6137964 F
6 G	99998771	10,6137179 F
7 G	99999834	12,6137073 F
8 G	99999980	14,6137059 F
9 G	99999997	16,6137057 F
10 G	99999999$\frac{3}{5}$	18,6137056 F

to obtain the order of accuracy of Newton. It boggles the mind that Newton would be carrying out such lengthy and tedious computation to unnecessary precision while writing his *Principia.* For comparison, from Eqs. (13) and (15) I have recalculated this table by computer (given below as Table 2); it shows that with few exceptions he obtained the correct answer up to round-off values in the last digit.

Table 2 Modern computation.

t/G	v/H	y/F
.1	.09966800	.00998338
.2	.19737534	.03973612
.3	.29131263	.08868153
.4	.37994894	.1559069
.5	.46211714	.2402290
.6	.53704959	.3402706
.7	.60436779	.45454043
.8	.66403675	.58150715
.9	.71629786	.71966094
1	.76159418	.86756164
2	.96402758	2.65000558
3	.99505478	4.61865711
4	.99932933	6.61437654
5	.99990922	8.61379623
6	.99998772	10.61371803
7	.99999833	12.61370754
8	.99999976	14.61370564
9	.99999994	16.61370659
10	1.00000000	18.61370659

Left: Lu Dalong (of Beijing); **Middle**: J. Fauvel; **Right**: I.B. Cohen

Newton's Mathematical Language

JOHN FAUVEL

Mathematics Department, Open University
Milton Keyes MK7 6AA, U.K.

He that would understand a book written in a strange language must first learn the language, and if he would understand it well must learn the language perfectly. Such a language was that wherein the Prophets wrote, and the want of sufficient skill in that language is the main reason why they are so little understood. John did not write in one language, Daniel in another, Isaiah in a third, but they all write in one and the same mystical language ... [which] as far as I can find, was as certain and definite in its signification as is the vulgar language of any nation.

Isaac Newton, *Theological Manuscripts*

Isaac Newton was one of the most self-conscious of mathematicians, deeply interested in questions of language, from a variety of perspectives, and deeply aware of the significance of linguistic and symbolic choices in the development of mathematics. Shortly after he went up to Cambridge in 1661, he indulged in the good seventeenth century occupation of constructing a universal language,[1] and his interest in language and symbolism continued throughout his creative life. The variety of evidence for Newton's fascination with symbolic representation ranges from his use of Shelton's shorthand system to record his adolescent sins[2] to his exploration of alchemical symbolism.[3] Such interests and concerns served a variety of purposes; most deeply it is clear that Newton believed the wisdom of the ancients had been encoded in symbols which the initiated could interpret and understand.[4] Again, this view is not uncharacteristic of a strong tradition found in that and the previous century, though it is fair to say that Newton went further than most in apparently attributing the inverse square law of gravitation to Pythagoras.[5]

In the seventeenth century, when algebraic symbolism was developing into so powerful and efficient an aspect of mathematical language, many drew parallels between mathematics and the possibility of a universal language. Robert Boyle, for example, wrote to Samuel Hartlib in 1647 "*Since our arithmetical characters are understood by all the nations of Europe, the same way, though every several people express that comprehension with its own particular language, I conceive no impossibility, that opposes the doing that in words, that we see already done in numbers;*" and William Oughtred's pupil Seth Ward wrote in 1654, "*An Universal Character might easily be made wherein all Nations might communicate together, just as they do in numbers and in species.*" Many more mathematicians of the period had deep interest in languages and linguistic notation, starting with Thomas Harriot, the greatest English mathematician before Newton, who in the 1580s was one of the early European explorers of North America, and the first person to take a formal linguistic interest in it, devising a notation for writing down the Algonquian language in Virginia.[6] And later in the seventeenth century, we have the well-known example of Leibniz's thoroughly mathematized *characteristica universalis.*[7] In his life-long concern with issues of language, Newton is to this extent a characteristic seventeenth century figure.

In his mathematical writings, Newton's conscious interest in language is often subsumed in the straightforward doing of mathematics, though here too an analysis of the different linguistic registers of his work can be very fruitful. A notable example of a very explicit attention to language is found in the text he deposited in the University Library some time after autumn 1683, purporting to be the algebra lectures he delivered as Lucasian professor. This contains a mass of material, from very simple instructions on the signs for addition and subtraction to novel and sophisticated results such as his extension of Descartes' rule of signs to imaginary roots, and was the basis for the *Arithmetica universalis* which appeared in 1707 edited by William Whiston, and in various guises throughout the eighteenth century so that it was eventually the most popular and widely disseminated of Newton's works, revealing a less familiar, or less well-remembered, side of Newton as a most skilful pedagogue. His views here date originally from 1670 when he was working, at John Collins' behest, on a project for introducing contemporary algebraic knowledge to the British reading public through an edition of a basic algebra text by the Dutch writer Gerard Kinckhuysen.[8] Newton took the opportunity to provide a thoughtful analysis of the transition in problem-solving from natural to mathematical language which shows the care and depth of his philosophical and surprisingly educational concerns.

> "The finding of equations — the part of this art which is by far the most difficult and yet is explained by no one — consists in devising an algebraic script by which the conditions of problems may be designated

> analogously to the way in which we denote our concepts in Greek or Roman characters. In this language, quantities fill the role of words, and equations that of sentences. Let quantities both known and unknown be designated by any symbols you wish as their algebraic words, and the sentences in which problems are enunciated will, when translated from Latin or any popular tongue into this language, then become the equations desired. And though the vernacular in which problems are propounded prove to be not at all suited to being written in algebraic form, yet by introducing appropriate changes and paying heed to the sense of words rather than to their spoken form, the change-over will be rendered an easy one. As a parallel, to be sure, the various national forms of speech have their own peculiar idioms and, when these are met with, translation from one to another is not to be instituted by a mere verbal transliteration, but must be determined from the sense. In the case of non-geometrical questions, however, if individual conditions are enunciated in distinct sentences or clauses of sentences, with parts, wholes and proportionals always designated by equalities between parts and wholes and between ratios or by effecting the multiplication of middles and extremes, I see nothing to block the path for beginners, on the pattern of the following examples."[9]

This passage resonates with Newton's other concerns of the period. For example, the very consideration of a method of finding equations reflects the issues around the invention of the calculus, in the balance between particular solutions of a problem and general analytical methods which might, in a particular case, be longer and less perspicuous, but are nevertheless better in the long run precisely on grounds of generality. In a letter to Collins of 27 September 1670, Newton commented on this balance in an educational context, in relation to the problems given in Kinckhuysen's book:

> "... having composed something pretty largely about reducing problems to an equation when I came to consider his examples [...] I found most of them solved not by any general Analytical method but by particular & contingent inventions, which though many times more concise than a general method would allow, yet in my judgment are less proper to instruct a learner, as Acrostick's & such kind of artificial Poetry though never so excellent would be but improper examples to instruct one that aim at Ovidian Poetry.[10]

In giving examples as Newton came to treat them, the clear and workable distinction between everyday language and its equivalent mathematical notation, amounting to a general translation method, is made very explicit through layout on the page as well as rhetorically.

A certain merchant each year increases his capital by a third, less £100 which he spends annually on his household, and after three years becomes twice as rich. What is his capital?

verbally	algebraically
A merchant has a certain capital:	x.
of this the first year he spends £100,	$x - 100$.
and increases the rest by a third;	$x - 100 + \frac{1}{3}(x - 100)$,
	or $\frac{1}{3}(4x - 400)$.
and the second year he spends £100,	$\frac{1}{3}(4 - 400) - 100$,
	or $\frac{1}{3}(4x - 700)$.
and increases the rest by a third;	$\frac{1}{3}(4x - 700) + \frac{1}{9}(4x - 700)$,
	or $\frac{1}{9}(16x - 2800)$.
and likewise the third year he spends £100,	$\frac{1}{9}(16x - 2800) - 100$,
	or $\frac{1}{9}(16x - 3700)$.
making a similar profit of one-third on the rest;	$\frac{1}{9}(16x - 3700) + \frac{1}{27}(16x - 3700)$,
	or $\frac{1}{27}(64x - 14\ 800)$.
and comes to be twice as rich as at the start.	$\frac{1}{27}(64x - 14\ 800) = 2x$.

"The question is accordingly reduced to the equation $\frac{1}{27}(64x - 14\ 800) = 2x$, and by its reduction x is to be determined. Namely, multiply it by 27 and it becomes $64x - 14\ 800 = 54x$; then subtract $54x$ and there remains $10x - 14\ 800 = 0$, that is, $10x = 14\ 800$. Therefore the initial capital, and also the profit, is £1480."

"You see, accordingly, that for the solution of questions whose preoccupation is merely with numbers, or the abstract relationships of quantities, almost nothing else is required than that a translation be made from the particular verbal language in which the problem is propounded into one (if I may call it so) which is algebraic; that is, into characters which are fit to symbolize our concepts regarding the relationships of quantities."[11]

Newton's explanation of the transition between natural and algebraic language is in a mathematical problem, and was later copied by textbook writers such as Lacroix.[12] It still makes good sense in today's classrooms.

When we turn to Newton's more famous and original mathematical work, the picture becomes rather more complicated. For someone with such an intense mathematical perception, and strong linguistic interests, he can be disconcertingly off-hand over notational matters to which you might (with hindsight) expect him to be more attentive. Take for example his early account of his fluxional researches *De methodis serierum et fluxionum*, written up in

1670–1671. Newton's Latin manuscript reads as follows

> "posthac denominabo fluentes, ac designabo finalibus literis v, x, y, et z. Et celeritates quibus singulae a motu generante fluunt et augentur (quas possim fluxiones vel simpliciter celeritates vocitare) designabo literis l, m, n et r. Nempe pro celeritate quantitatis v ponam l et sic pro celeritatibus aliarum quantitatum x, y, et z ponam m, n, et r respectivè."

D.T. Whiteside's translation of this passage reads:

> "I will hereafter call them fluents and designate them by the final letters v, x, y and z. And the speeds with which they each flow and are increased by their generating motion (which I might more readily call fluxions or simply speeds) I will designate by the letters $\dot{v}, \dot{x}, \dot{y}$ and $\dot{z}$: namely, for the speed of the quantity v I shall put $\dot{v}$, and so for the speeds of the other quantities I shall put $\dot{x}$, $\dot{y}$ and $\dot{z}$ respectively."[13]

This area of Newton's work is rather interesting for the relation between language, notation and thought. Whiteside commented in a footnote that he had translated "l" by "v-dot", etc. and thus rendered the original form in the dotted notation which was not in fact introduced by Newton for another twenty years (in late 1691) "*merely as an aid to comprehension of involved arguments not easy to follow in their original dress.*" Whiteside went on to observe that all four eighteenth century editors of the text introduced the dotted notation without comment, and that "*this practice has had its effect in seriously distorting ... all subsequent assessment of the relative merit of Newton's and Leibniz's early calculus symbolisms.*"

In short, Newton's early notation for the calculus, for its first twenty years and more, lacks precisely the characteristic notational feature that we expect: a symbolic connection between a variable and its differential, the connection that enables us to turn the handle of the great calculus sausage machine and see the results flow out automatically. To add further complications to the historic record, or reader's expectations, Whiteside's study of the early papers has shown that back in 1665, Newton did indeed briefly toy with a "double-dotted" notation for the derivative,[14] before abandoning it in favor of different letters. So, he certainly contemplated a notational link. And once he began using his new dotted notation (in 1692), within months it appeared in print through the agency of John Wallis, in the Latin edition of Wallis's *Algebra*,[15] and soon became established as *the* Newtonian notation for fluxions, with the general assumption that this was how he had originally conceived of it. In view of this, one may consider that Newton's understanding seems to have operated at a level where the notation wasn't an especially significant feature of the mathematics. Although for most of us it is the symbolism — particularly, now, the Leibnizian symbolism — of the calculus that paves the way for secure handling of the concepts and techniques, at another level the notation may well not be the key intuition involved.

This raises the question of Newton's mental processes and attributes, the gap between internal mental processes and the public language of symbolic communication. There is widespread agreement that one of the prime factors in Newton's achievement was his ability to hold a problem in his mind and focus on it over a long period of time. This classic Newtonian myth (which is not to say it is untrue) was notably expressed in the words of John Maynard Keynes, composed for the 1942 tercentenary celebrations:

> "I believe that the clue to his mind is to be found in his unusual powers of continuous concentrated introspection. ... His peculiar gift was the power of holding continuously in his mind a purely mental problem until he had seen straight through it. I fancy his pre-eminence is due to his muscles of intuition being the strongest and most enduring with which a man has ever been gifted. ... I believe that Newton could hold a problem in his mind for hours and days and weeks until it surrendered to him its secret. Then being a supreme mathematical technician he could dress it up, how you will, for purposes of exposition ... The proofs, for what they are worth, were, as I have said, dressed up afterwards — they were not the instrument of discovery.... Certainly there can be no doubt that the peculiar geometrical form in which the exposition of the *Principia* is dressed up bears no resemblance at all to the mental processes by which Newton actually arrived at his conclusions."[16]

This is a rich and provocative text. The contrast between contexts of discovery and of justification is one whose validity continues to be discussed today. There are particular issues concerning *Principia* to be discussed in a moment, but there are more general issues to look at first. Let us put the problem naively: what language did Newton use to think in mathematically? What mathematical language does *anyone* use? And how can we tell? These rhetorical questions help to clarify the point, that we can only really be concerned with *communication through* language; someone's self-communing, at the level of Newton's mathematical thought, is beyond language. In a sense, the claim that the way *Principia* was written down "*bears no resemblance at all*" to the mental processes of discovery is a truism, a remark with no content because it is self-evidently and almost analytically true. As a matter of fact, it seems that

Newton made his discoveries much more slowly and painfully than the myth would have it: "The record shows that he made mistakes, that he learned from them, and that with unwearying application he steadily enlarged his grasp as he constructed the mature fluxional calculus."[17]

One celebrated issue that it is useful to spend a little time on is the question of the language of *Principia.* The story is still going around — it is certainly hinted at in Keynes's remarks — that Newton worked out the results of *Principia* in fluxions, in the language of the calculus, and *then translated or dressed up* his work into the language of geometry for some reason. This story dies hard, partly because Newton himself gave it currency, for reasons of his own, although D.T. Whiteside has been pointing out for many years its lack of any evidential basis. Exploring the complexities of the question is quite revealing, however. For one thing, the sense in which *Principia* is indeed geometrical needs careful qualification. Although it is obviously in a Euclidean *form* of definition, axiom, theorem and proof, the mathematics funnelled into this conventional form is by no means classical geometry: it's not something which either looks or reads, except sporadically, like Euclid or Apollonius. The geometry involved is late seventeenth century geometry: Newtonian geometry, in fact. Secondly, this geometry is indeed already a kind of geometrized infinitesimal calculus; there's no need to postulate that it is translated *from* the calculus when it is evidently imbued with that spirit and type of argument.

This can be seen by taking any example: the very first theorem of *Principia* is a good case in point. This is the theorem in which Newton showed that the area swept out by the radius of a body moving in a central-force field is proportional to the orbital time (Kepler's second law, in other words, extended to an arbitrary situation in which a force is directed towards a fixed point). Here, the method of proof is by supposing the body moves along a series of infinitesimal straight lines, in such a way that it can be considered as moving ahead in a straight line and then knocked by the impulse of the force back towards the center. The argument then depends on a careful choice of what happens when the infinitesimals tend to zero. This is quite recognizable as a geometrized calculus, albeit without its algorithmic aspect. It is a long way from classical geometry, being indeed a complex interplay of geometry and physics. It is quite different, for example, from the work in "pure" classical geometrical analysis, following Euclid, Apollonius and Pappus, which Newton carried out a few years later. Furthermore, as Emily Grosholz has shown, the diagrams in *Principia* play a role so crucial as to form a second, iconic, linguistic dimension to complement the role of symbolic language in the prose rhetoric of a proposition: the diagrams, such as that for Theorem 1, are essential to interpreting the text, and provide further evidence of the un-Euclidean nature of the geometry.[18]

In the preface to *Principia*, Newton explained his view of how geometry relates to natural philosophy: "geometry is founded in mechanical practice, and is nothing but that part of universal mechanics which accurately proposes and demonstrates the art of measuring." He then promoted the view that mathematics is the name of the kind of reasoning that he is using to lead from physical, mechanical principles to conclusions about the world:

> "... we offer this work as mathematical principles of philosophy, for all the difficulty of philosophy seems to consist in this — from the phenomena of motions to investigate the forces of nature, ... In the third book we give an example of this in the explication of the System of the World; for *by the propositions mathematically demonstrated* in the first book, we then derive from the celestial phenomena the forces of gravity with which bodies tend to the sun and the several planets. Then from these forces, *by other propositions which are also mathematical*, we deduce the motions of the planets, the comets, the moon and the sea. I wish we could derive the rest of the phenomena of nature by the same kind of reasoning from mechanical principles ... [emphasis added]."

Notice the way Newton emphasized the mathematical nature of the demonstrations, as though there were other styles of argument available but which he is rejecting. This of course is the essence of the Newtonian revolution in science.[19] Some indication of the case that Newton was implicitly arguing against may be gathered from, for example, Robert Boyle's doubts about the appropriateness of mathematics within natural philosophy.[20]

Despite one's qualifications on careful examination of *Principia*, it was certainly the Euclidean form of *Principia* which impressed several of his contemporaries, and for whom this aspect accounted for the book's success. John Locke was one such reader of *Principia*, and a most influential one. Locke emphasized in the detailed review he gave the book that the key to Newton's success was the method of geometry, the movement from axioms and definitions to conclusions through rigorous intermediate steps.[21] In his "*Essay concerning human understanding*," Locke emphasized as tools of philosophical inquiry, both the geometrical style of deduction which he had seen so effective in *Principia*, and the algebraic language which he was also highly impressed by — but he was aware, too, of semeiotic differences between mathematical and moral sciences: "*Indeed, wrong Names in moral discourses, breed usually more disorder, because they are not so easily rectified, as in Mathematicks, where the Figure once drawn and seen, makes the name useless and of no force. For what need of a Sign, when the Thing signified is present and in view? But in moral Names, that cannot be so easily and shortly done,...*"[22]

As a cheerful example of how Newton's formal influence could operate, we may look briefly at John Craig's *Theologiae christianae principia mathematica* (1699). This theological work, an attempt to calculate the date of Christ's second coming, was closely modeled on the formal style and structure of Newton's Principia, with definitions, three laws, and theorems such as "*Velocities of suspicion produced in equal periods of time increase in arithmetical progression.*" Craig was a Scottish mathematician who knew Newton and was one of the early writers on the fluxional calculus. It is fair to add that Craig's work did not meet with the same success as Newton's, which may suggest that the formal geometrical appearance of a work is not the sole determinant of its persuasive qualities.

There is another story about the early calculus which casts light on our theme, besides the story about *Principia* being worked out by fluxional calculus and then translated into a different language. This other story is that the fundamental theorem of the calculus was in fact known to Newton's patron and predecessor as Lucasian professor, Isaac Barrow. Again it is revealing, on several levels, to unravel how such a misperception could arise. It is true that Proposition x : 11 of Barrow's *Lectiones geometriae*[23] shows that a line constructed to equal (in some sense) the area under the curve has sub-tangents relating to the original ordinates, and that this result can be read, with hindsight, as showing that integration and differentiation are converse operations. But there are conceptual and historiographical objections to such a reading. One is that nobody at the time, not even Barrow, attached any such significance to it; another that Barrow's theorem was in any case basically a restyling of James Gregorie's generalization of William Neile's rectification method. More deeply, though, there is a whole operational, algorithmic, generalizable dimension to the calculus which Barrow did not have in mind; intention is important here. As the historian Michael Mahoney has judiciously summarized it, after tracing the way that the calculus of both Newton and Leibniz relied fundamentally on algebraic analysis: "Thinking algebraically counted. That is why historians of mathematics must be careful about referring to Barrow's methods as calculus in another — in geometric — disguise. Conceptually the guise is everything."[24]

At the same time, Newton did not in this case play down his inspirational debt to Barrow, writing in 1712 that it was from Barrow

> "that I had the language of momenta & incrementa momentanea & this language I have always used & still use as may be seen in my Analysis per aequationes infinitas communicated by Dr Barrow to Mr Collins A.C. 1669 & in my Principia Mathematica & Quadratura Curvarum. And by putting the velocities of the increase of quantities proportional to the incrementa momentanea I found out the demonstration of the method of moments & thence called it the method of fluxions."[25]

The appropriate kind of mathematical language, in particular the roles of algebra and geometry, was a long-standing concern of Newton's, preserved in several stories emanating from his circle. David Gregory recorded that Newton once said "*Algebra is the Analysis of the Bunglers in Mathematicks*," while Henry Pemberton reported the strength of Newton's views on mixing algebra and geometry: "*I have often heard him censure the handling [of] geometrical subjects by algebraic calculations; and his book on Algebra he called by the name of Universal Arithmetic, in opposition to the injudicious title of Geometry, which Des Cartes had given to the treatise wherein he shows, how the geometer may assist his invention by such kind of computations.*"[26] Indeed, in the deposited lectures on algebra, Newton wrote

> "Equations are expressions belonging to arithmetical computation and in geometry properly have no place except in so far as certain truly geometrical quantities (lines, surfaces and solids, that is, and their ratios) are stated to be equal to others. Multiplications, divisions and computations of that sort have recently been introduced into geometry, but the step is ill-considered and contrary to the original intentions of this science. ... Consequently these two sciences ought not to be confused. The Ancients so assiduously distinguished them one from the other that they never introduced arithmetical terms into geometry; while recent people, by confusing both, have lost the simplicity in which all elegance in geometry consists."[27]

Although not published until 1707, these sentiments date from the mid-1680s, some twenty years after his first introduction to Descartes' work. The whole question of the legitimacy of admitting different techniques or styles into parts of mathematics is a profound concern throughout the seventeenth century, certainly since Descartes raised the question of using algebra in solving geometrical problems. Newton made one of the most serious studies of this question, over a period of some forty years, and came to take as we have seen quite a conservative position: that is, while conceding that algebra is useful as a tool, he felt it to be a bad guide in questions concerning the aims and methods of geometry. This is but a particular case of a concern, from Descartes himself onwards, for what was or was not properly mathematical. It can be a little hard for us now to see what the problem was, but it was felt by other mathematicians too, as well as Newton, that geometry had its own programme and unbridled use of alien techniques would be ungeometrical.

Ironically, the calculus itself was viewed somewhat ambivalently, in this alien light, by at least one leading mathematician of the older generation, Christiaan Huygens. Huygens' point was that the algorithmic, handle-turning aspect of the calculus produces an answer, a number or whatever, but in the process deprives you of *understanding* the geometry of the situation.[28] This argument still has considerable force.[29]

The alert reader of the above passage from Newton will at once notice a none-too-disguised belittling of Descartes, to whom Newton nevertheless owed so much. Whiteside's judgement is clear that the decisive influence enabling Newton to have so versatile and varied a set of mathematical techniques at his disposal was Descartes' *La geometrie*, which Newton read in van Schooten's Latin edition in 1664.

> "What appears to me ever more clearly to be the decisive event guiding the future direction of Newton's mathematical thought and practice was his reading of Descartes' *Geometrie* in the late summer of 1664. ... Above all, I would assert, the *Geometrie* gave him his first true vision of the universalising power of the algebraic free variable, of its capacity to generalise the particular and lay bare its inner structure to outward inspection."[30]

It was the revelation of what was possible through symbolic mathematics, the stretching of his imagination and conceptions, that seems to have opened up Newton's latent power, consolidating the introduction to symbolism which he had received a few months before from studying William Oughtred's *Clavis mathematicae*. The fact that he spent much of his later career rejecting various parts of the Cartesian legacy adds to the piquancy of Newton's complicated relationship with his precursors.

I end with an example to illustrate the illumination we can glean about the mathematics developing under Newton's care in the late 1660s. One of the most celebrated of his manuscripts, the *De analysi* of around 1669, contains an account of his use of infinite series in geometrical analysis, produced we may suppose in response to Newton's realizing from the publication of Nicholas Mercator's *Logarithmotechnia*, the previous year, that Mercator and others were hard on his heels.

This manuscript starts with the three rules for integrating curves, in effect, which he had discovered in the year or so up to 1666. First, if the curve is $y = ax^{m/n}$, then the area under it is given by the usual rule for powers of x; second, the addition rule, that you can integrate each term of a sum separately and add the results; and thirdly, that if the formula for y is a bit more complicated you must reduce it to simpler terms "*by operating in general variables in the same way as arithmeticians in decimal numbers divide, extract roots or solve affected equations.*" And he then went on to do precisely that, to exemplify these three ways of reducing an expression to a state where you can handle it more easily, in an infinite series expansion which you can integrate term by term.

The important thing to notice is the clear overt way in which Newton treated algebraic expressions as though they were calculations in decimal arithmetic. At one level it is quite astonishing in its simplicity and straightforwardness. Of course, it is the most simple mathematics which requires genius to uncover. We know that the developing infinitesimal calculus made extensive use of Newton's technique of infinite series. And the algebra involved in that is based in turn simply on calculations in decimal arithmetic such as one might have learned at school. One might want to go further with speculations about the significance of decimal calculations on the use and spread of algebra.

If this seems implausible at first, one should look at the method Newton employed in his next example, where he simplified $y = \surd(a^2+x^2)$ and obtained $a + x^2/2a - x^4/8a^3 + \ldots$ Again he did this by simply applying a technique of schoolroom mathematics (something known for generations, if not millennia, from the Yellow River to the Straits of Gibraltar), namely the extraction of a square root — but applying it to a quite novel object — an algebraic expression. We can read how to extract roots in his *Universal arithmetic*, or in the lectures he deposited in the University Library,[32] and indeed in any competent arithmetic text since the invention of printing.

The third example in *De analysi* is also rather interesting, the "numerical resolution of affected equations." This is probably the most familiar of the three techniques now, because it looks suspiciously like the "Newton–Raphson method" for deriving the root of an equation, which has sprung back into prominence because of its utility in electronic calculators and computers. Was it discovered by Newton, and if so what did Raphson contribute? Newton himself commented "*I do not know whether this method of resolving equations is widely known or not.*" D.T. Whiteside's characteristically invaluable note points out in response that the method "*has a long manuscript history, and its essential structure was known to ... al-Kasi.*" But it is another matter altogether whether what appears in *De analysi* is what is now understood by the Newton-Raphson method, namely an iterative approximation technique employing the calculus. Recently, Nick Kollerstrom has shown that in the latter sense the method is due to neither Newton nor Raphson, but to Thomas Simpson in 1740, some seventy years later.[33] Kollerstrom's work is a good example of the further advances in historical research that the great editions of Newton's works produced in recent years have made possible.

Notes and References

1. Ralph W.V. Elliott, "Isaac Newton's 'Of an universal language'," *Modern Language Review* **52** (1957): 1–18.
2. Richard S. Westfall, "Short-writing and the state of Newton's conscience, 1662," *Notes and Records of the Royal Society of London* **18** (1963): 10–16.
3. Jan Golinski, "The secret life of an alchemist," in *Let Newton Be!* (ed.) John Fauvel *et al.* (Oxford University Press, 1988), pp. 147–167.
4. J.E. McGuire and P.M. Rattansi, "Newton and the 'pipes of Pan'," *Notes and Records of the Royal Society of London* **21** (1966): 108–143.
5. Piyo Rattansi, "Newton and the wisdom of the ancients" in *Let Newton be!* (ed.) John Fauvel *et al.* (Oxford University Press, 1988), pp. 185–201.
6. Vivian Salmon, *Thomas Harriot and the English Origin of Algonkian Linguistics*, Durham Thomas Harriot Seminar Occasional Paper #8 (1993).
7. Joseph E. Hofmann, *Leibniz in Paris 1672–1676: His Growth to Mathematical Maturity* (Cambridge University Press, 1974).
8. C.J. Scriba, "Mercator's Kinckhuysen-translation in the Bodleian Library at Oxford," *British Journal for the History of Science* **2** (1964): 45–58.
9. Isaac Newton, "*First book of universal arithmetic*," manuscript of c.1684, in *The Mathematical Papers of Isaac Newton* (ed.) D.T. Whiteside (Cambridge, Cambridge University Press, 1972), Vol. 5, pp. 565–569.
10. Newton to Collins, 27 September 1670, in *The Correspondence of Isaac Newton* (ed.) H.W. Turnbull (Cambridge, 1959), Vol. 1, pp. 43–44.
11. D.T. Whiteside, *The Mathematical Papers of Isaac Newton* (Cambridge, Cambridge University Press, 1972), Vol. 5, p. 133.

12. Henry Plane, "Translating into algebra," in *History in the Mathematics Classroom: The IREM Papers* (ed.) John Fauvel (The Mathematical Association, 1990), pp. 59–72.
13. D.T. Whiteside, *The Mathematical Papers of Isaac Newton* (Cambridge, Cambridge University Press, 1969), Vol. 3, p. 73.
14. D.T. Whiteside, *The Mathematical Papers of Isaac Newton* (Cambridge, Cambridge University Press, 1967), Vol. 1, p. 363.
15. John Wallis, *De Algebra Tractatus; Historicus & Practicus* (Oxford, 1693), pp. 390–396.
16. J. Maynard Keynes, "Newton, the man," *Essays in Biography* (Mercury Books, 1961), pp. 310–323.
17. Richard S. Westfall, "Newton's marvelous years of discovery and their aftermath: myth versus manuscript," *Isis* **71** (1980): 109–121: 113.
18. Emily Grosholz, "Some uses of proportion in Newton's *Principia*, book I: a case study in applied mathematics," *Studies in the History and Philosophy of Science* **18** (1987): 209–220.
19. I. Bernard Cohen, *Revolution in Science* (Harvard University Press, 1985).
20. Steven Shapin, "Robert Boyle and mathematics: reality, representation, and experimental practice," *Science in Context* **2** (1988): 23–58.
21. James L. Axtell, "Locke's review of the *Principia*," *Notes and Records of the Royal Society of London* **20** (1965): 152–161.
22. John Locke, *An Essay Concerning Human Understanding* (ed.) Peter H. Nidditch (Oxford University Press, 1975).
23. D.J. Struik, *A Source Book in Mathematics*, 1200–1800 (Harvard University Press, 1969), pp. 255–256.
24. Michael Mahoney, "Barrow's mathematics: between ancients and moderns," in *Before Newton: The Life and Times of Isaac Barrow* (ed.) M. Feingold (Cambridge University Press, 1990), 179–249: 237.
25. A. Rupert Hall and Laura Tilling (eds.), *The Correspondence of Isaac Newton* (Cambridge, 1975), Vol. 5, p. 213.
26. I. Bernard Cohen, entry on Newton in *Dictionary of Scientific Biography* (ed.) C.C. Gillispie (Charles Scribner's Sons, 1974), Vol 10, 42–101: 88.
27. D.T. Whiteside, *The Mathematical Papers of Isaac Newton* (Cambridge, Cambridge University Press, 1972), Vol. 5, p. 429.
28. Bos, H.J.M., "Huygens and mathematics," in *Studies on Christiaan Huygens* (ed.) H.J.M. Bos *et al.* (Lisse, 1980), pp. 126–146.

29. Jan van Maanen, "From quadrature to integration: thirteen years in the life of the cissoid," *Mathematical Gazette* **75** (1991): 1–15.
30. D.T. Whiteside, "Newton the mathematician," in *Contemporary Newtonian Research* (ed.) Z. Bechler, 1982, 109–127: 113–114.
31. D.T. Whiteside, *The Mathematical Papers of Isaac Newton* (Cambridge, Cambridge University Press, 1968), Vol. 2, p. 213.
32. D.T. Whiteside, *The Mathematical Papers of Isaac Newton* (Cambridge, Cambridge University Press, 1972), Vol. 5, pp. 89–101.
33. Nick Kollerstrom, "Thomas Simpson and 'Newton's method of approximation': an enduring myth," *British Journal for the History of Science* **25** (1992): 347–354.

Newton's Expansion for the Square Root of an Algebraic Equation by an Equivalent Arithmetic Method

MICHAEL NAUENBERG

Department of Physics, University of California
Santa Cruz CA 95064, U.S.A.

In the article titled "Newton's Mathematical Language," Chapter 9, Fauvel discussed Newton's remarkable expansion of certain algebraic expressions "as though they were calculations done in decimal arithmetic." In particular, Newton developed the infinite series expansion of $\sqrt{a^2+x^2}$ by applying directly an arithmetical technique for evaluating the square root of ordinary numbers.[1] This technique used to be taught (by rote) in schools, but this is now rarely the case. I present here a justification for Newton's method, and correspondingly for the arithmetic technique, which is usually not given in the textbooks, and also show the relation of this technique to the better known Newton–Raphson method.[2]

Newton's fundamental idea for the solution of general algebraic equations was to generate successive improved approximations by treating certain terms as small quantities. This is illustrated here by assuming that x is much smaller than a and approximating $\sqrt{a^2+x^2}$ by a sequence $a_n + b_n$, treating b_n as a small quantity of order x^{2n}. Then

$$(a_n + b_n)^2 = a_n^2 + 2a_n b_n + b_n^2 \approx a^2 + x^2 \tag{1}$$

where $a_{n+1} = a_n + b_n$ and $a_1 = a$. Keeping only terms up to order x^{2n}, the term b_n^2 can be neglected and $2a_n b_n$ replaced by $2ab_n$. Hence

$$b_n = \frac{(a^2 + x^2 - a_n^2)}{2a}, \tag{2}$$

where only the lowest order term (order x^{2n}) on the right-hand-side is retained. The recurrence relation for $a_{n+1} = a_n + b_n$ can be applied to simplify the calculation of this equation by setting the remainder in the form

$$a^2 + x^2 - a_{n+1}^2 = (a^2 + x^2 - a_n^2) - (2a_n + b_n)b_n\,. \tag{3}$$

This procedure corresponds to the arithmetic routine for evaluating square roots which is reduced here to the recursive computations of $(2a_n + b_n)b_n$ subtracted from the remainder $a^2 + x^2 - a_n^2$.

The first step, $n = 1$, sets $a_1 = a$, and

$$a^2 + x^2 - a_1^2 = x^2 \,. \tag{4}$$

Hence, according to Eq. (2)

$$b_1 = \frac{x^2}{2a} \,, \tag{5}$$

which gives the first approximation to $\sqrt{a^2 + x^2}$,

$$a_2 = a_1 + b_1 = a + x^2/2a \,. \tag{6}$$

For $n = 2$, we have

$$a^2 + x^2 - a_2^2 = x^2 - (2a + b_1)b_1 = -\frac{x^4}{4a^2} \,, \tag{7}$$

substituting in Eq. (2) gives

$$b_2 = -\frac{x^4}{8a^3} \,, \tag{8}$$

and the next order term in the expansion,

$$a_3 = a_2 + b_2 = a + \frac{x^2}{2a} - \frac{x^4}{8a^3} \,. \tag{9}$$

The next order, $n = 3$, illustrates the procedure further. We have

$$a^2 + x^2 - a_3^2 = -\frac{x^4}{4a^2} - (2a_2 + b_2)b_2 = -\frac{x^6}{8a^4} + \frac{x^8}{64a^6} \,, \tag{10}$$

but when substituting this expression in Eq. (2) to evaluate b_3, one keeps only the first term (order x^6),

$$b_3 = -\frac{x^6}{16a^5} \,. \tag{11}$$

However, both terms in Eq. (10) must be included in the evaluation of the next difference $a^2 + x^2 - a_4^2$.

It can be readily verified that this procedure gives the same terms $a^2 + x^2 - a_n^2$ which Newton obtained as intermediary steps up to order $n = 7$ by applying the rules for the arithmetic extraction of square roots.[1] The validity for this analogy depends of course on the fact that the arithmetical rules are justified by a very similar procedure as we will now show briefly.

Let z be a number given by a decimal expansion. Then its square root $\sqrt{z}$ is approximated by the sequence

$$\sqrt{z} \approx a_n + \frac{b_n}{10^n} \tag{12}$$

where b_n are integers in the range $0 - 9$ and $a_{n+1} = a_n + b_n/10^n$. Here, a_1 is the integer whose square is the closest approximate to z for which $a_1^2 < z$ and b_n is the integer such that $2a_nb_n$ best approximates but is smaller than $(z - a_n^2)10^n$. As before, the difference or remainder $z - a_n^2$ is evaluated by the recurrence relation

$$z - a_{n+1}^2 = z - a_n^2 - (2a_n + b_n/10^n)(b_n/10^n) \tag{13}$$

which must be evaluated exactly to $2n$ decimal places by carrying out the sum and multiplication indicated in the last term.

Fauvel commented that Newton also developed a method which looks "suspiciously" like the "Newton–Raphson method" for extracting square roots. In fact, this method corresponds to setting

$$b_n = \frac{z - a_n^2}{2a_n}, \tag{14}$$

and the recurrence relation for a_n gives

$$a_{n+1} = \frac{1}{2}\left(a_n + \frac{z}{a_n}\right). \tag{15}$$

This method converges much more rapidly, but it requires a division in the second term, while the previous method is based on using only sums, multiplications and subtractions.

Acknowledgments

I would like to thank Professor David Fowler for some very interesting discussions on the subject of this note.

Notes and References

1. *The Mathematical Papers of Isaac Newton 1670–1673* (ed.) D.T. Whiteside (Cambridge University Press, 1969), Vol. 3, p. 41. In this example, Newton's calculation for the first seven terms of this expansion is organized in the same manner as in the corresponding school method for extracting square roots.
2. N. Kollerstrom recently argued that the attribution of this method to Newton is an "enduring myth", evidently failing to understand the equivalence of Newton's method with the modern formulation. See "Thomas Simpson and 'Newton's method of approximation': an enduring myth," *British Journal for the History of Science* **28** (1992): 347–354.

Bernard Cohen (right) and Michael Nauenberg (left).

Newton's Portsmouth Perturbation Method and its Application to Lunar Motion

MICHAEL NAUENBERG

Department of Physics, University of California
Santa Cruz CA 95064, U.S.A.

Abstract

The Portsmouth collection of Newton's mathematical papers shows that by 1687, Newton had developed a general perturbation method to deal with the effect of solar gravitation on the motion of the moon. However, this method did not appear in the *Principia* except for a brief outline in Corollaries 3 and 4 of Proposition 17 of Book I. It will be shown that this perturbation method corresponds to the method of variation of orbital parameters developed later by Euler, Lagrange and Laplace. Newton obtained an equation for the rotation and oscillation of the lunar apsides which provides the gravitational basis for Horrock's kinematical model of the evection. Newton's integration of this equation is discussed and compared with the results obtained by Clairaut, d'Alembert.

1 Introduction

One of Newton's mathematical manuscripts in the Portsmouth collection, first examined in 1872 by a syndicate appointed by the University of Cambridge,[1] indicates that by 1687 Newton had developed a general method to deal with the effect of perturbations on the Keplerian motion of planets. The propositions and lemmas in this manuscript where apparently intended for inclusion in the *Principia*, but these did not appear in any one of the three editions supervised by Newton. However, a brief outline can be found in the first edition, Corollaries 3 and 4 of Proposition 17, although the connection of these corollaries to the Portsmouth manuscript has not been realized previously.[2] The only application of this method which appears in these manuscripts is a calculation of the mean rotation of the lunar apsis due to the perturbation of the sun, which suggests that Newton had developed this method specially for this purpose. Previously he had developed an approximate perturbation method discussed in Sec. 9 of the *Principia* which had failed to account quantitatively for the observed mean rotation by a factor of two. However, by 1687 Newton appeared to believe that he had succeeded in calculating the motion of the lunar apogee in agreement with observation, as he announced in the scholium to Proposition 35, Book III in the first edition of the *Principia*:

> "Thus far regarding the motions of the moon insofar as the eccentricity of the orbit is not considered. By similar computations I have found that the apogee, when it is situated in conjunction or opposition to the sun, advances each day 23′ in respect to the fixed stars. but at the quadratures regresses each day about 16 1/3′; and that its mean annual motion should be virtually 40°. By the astronomical tables adapted by Mr. Flamsteed to Horrocks hypothesis the apogee at its syzygies advances with a daily motion of 20′ 12″, and is borne in *consequentia* with a mean annual motion of 40° 41′ ... These computations, however, excessively complicated and clogged with approximations as they are, and insufficiently accurate we have not seen fit to set out [*computationes autem, ut nimis perplexas & approximationibus impeditas, neque satis accuratas, apponere non lubet*],"[3]

However, in the two subsequent editions of the *Principia*, this scholium was revised leaving out these claims entirely,[4] and in the third edition he inserted the well-known remark at the end of Corollary 2, Proposition 45 of Sec. 9 that "the apse of the moon is about twice as swift" as the value calculated by the perturbation method in this section.

Newton's perturbation method in the Portsmouth manuscript is based on a geometrical approach which leads directly to differential equations, which he referred to as *hourly motion*, for the parameters of an ellipse describing orbital motion due to an inverse square force in the presence of an external perturbation force. It will be shown that this method corresponds to the modern method of variation of orbital parameters attributed to Euler, Lagrange and Laplace,[5–8] although this connection has not been generally appreciated in the past.[9–15] In 1894, the French astronomer Tisserand discussed Newton's treatment of the lunar problem in his treatise entitled "Celestial Mechanics,"[16] and included a brief description of a few of the equations obtained in the Portsmouth manuscripts comparing them with the results of Euler and Laplace. However, no further comments of this connection appeared in the literature, and when these manuscripts were published by D.T. Whiteside in 1974, with extensive footnotes in Vol. 6 of Newton's *Mathematical Papers*,[17] no indication was given of the relation of Newton's method to the modern variation of orbital parameters method. Actually, in the Portsmouth manuscripts, one finds only the variational equation for the rotation of the axis of the ellipse, while the corresponding equation for the eccentricity is absent. Moreover, the variational equation is obtained only to lowest order in the eccentricity. Integrating this equation, Newton obtained a correction to the rate of rotation of the apsis which he had previously calculated in Proposition 45. His result is comparable to the correct value obtained much later by Clairaut and d'Alembert[18] who applied a different approach (see Appendix A). It will be shown that the difference can be traced to some errors in Newton's expansion of his equation, and to his neglect of certain higher order terms in the eccentricity of the ellipse. Moreover, Newton's equation implies oscillations of the lunar apsides which correspond to those in Jeremy Horrock's kinematical model, developed to account for the *evection.*

2 Portsmouth Perturbation Method

The starting point of Newton's Portsmouth method is to replace a continuous perturbing force acting on a body revolving in an orbit under the action of an inverse square force by a discrete sequence of instantaneous impulses at short equal time intervals. Then, in between impulses, a segment of the orbit corresponds to the arc of an ellipse with a focus fixed at the center of the inverse square force. According to Proposition 17 in Book I of the *Principia,*[21] the parameters of this ellipse (semi-latus rectum, eccentricity, and spatial orientation) are determined *uniquely* from the velocity and position of the body at any instance of time.[19] These parameters are evaluated after the action of each of these perturbing impulses which change *instantaneously* the magnitude and direction of the velocity of the body without changing its position. In Corollary 3 of Proposition 17, Newton outlined this perturbation method rather succinctly in the following form:

> "Hence also, if a body moves in any conic whatever and is forced out of its orbit by any impulse, the orbit in which it will afterwards pursue its course can be found. For by compounding the body's own motion with that motion which the impulse alone would generate, there will be found the motion with which the body will go forth from the given place of impulse along a straight line given in position."[20]

Repeated applications of Proposition 17 then determines the perturbed orbit of the moon as a sequence of elliptical arc segments joined together. As Newton explained in Corollary 4,

> "And if the body is continually perturbed by some force impressed from outside, its trajectory can be determined very nearly by noting the changes which the force introduces at certain points and estimating

> from the order of the sequence the continual changes at intermediate places.[20]"

However in the *Principia*, Newton did not give any further details of his perturbation method. In the Portsmouth manuscript, he stated more precisely how to calculate the perturbed orbit,

> "... conceive now that the impulses are increased in number, and their intervals diminished indefinitely so as to render the action of the forces ... continuous."[17]

Taking the limit of a vanishingly small time interval between impulses leads to differential equations, which Newton called *hourly motion*, for the evolution of the parameters of the ellipse which determines locally the orbit. Newton's Portsmouth method corresponds[9] to what is now known as the method of variation of orbital parameters generally attributed to Euler, Lagrange and Laplace who developed this method many years later.[5–8] In the Portsmouth manuscript, one finds only the derivation of the differential equations for the rotation of the axis of this ellipse, evaluated to lowest order in the eccentricity e, due to a radial and a transverse perturbation force. Newton integrated this equation for the case of the moon orbiting the earth where the main perturbation is due to the gravitational attraction of the sun, which he determined to first order in the ratio of the earth-moon and earth-sun distances.

In a similar manner, Newton could have obtained also the corresponding change in the eccentricity e and the magnitude of the major axis of the ellipse, but there is no evidence in the Portsmouth manuscripts that he carried out these calculations. However in Corollaries of Proposition 66, he discussed the variations of the eccentricity qualitatively, and in the revised scholium after Proposition 35, Book III, which appeared in the second and third editions of the *Principia*, he discussed both the variation of the eccentricity as well as the rotation and variation of the the moon's apsis in connection with a model which he attributed to Horrocks and to Halley.[22] In this model, the moon rotates on an ellipse with the earth at one of the foci, while the center of this ellipse rotates on an epicycle around the earth. There is much confusion in the literature on the relation between Horrocks kinematical model and Newton's gravitational theory. It is generally thought that Newton had not been able to deduce this model from his theory of gravity (see for example, Whiteside Ref. 3), although he had claimed in this scholium that it followed "by the same theory of gravity." It will be shown here that to first order in the eccentricity this model is indeed obtained by solving Newton's perturbation equation, although the magnitude for the perturbing parameter is somewhat different from the value given in the revised scholium to Proposition 35.

The Portsmouth manuscript starts with two lemmas where Newton derived separately the effect of a radial perturbing force V and a transverse perturbing force W by applying Proposition 17. In this proposition, Newton demonstrated how the parameters of a conic section orbit for an inverse square force are determined by the position and velocity of the orbiting body at a given instant of time. Referring to Fig. 1 for an ellipse with foci at S and F, suppose the center of force is at S, the position of the body is at P, and the tangent line Pp gives the direction of the velocity v of the body at P. From the properties of an ellipse, we have $SP + PF = QR$, where $QR/2$ is the major axis of the ellipse, and $SF^2 = QR^2 - 4SP \times PF \times \sin^2(\alpha)$, where SF is the distance between the foci, and α is the angle between SP and the tangent Pp. Setting $L = (QR^2 - SF^2)/QR$, where L is the latus rectum of the ellipse,

$$PF = \frac{L \times SP}{4SP \sin^2 \alpha - L} \tag{1}$$

which is the basic relation of Proposition 17. The latus rectum L is proportional to the square of the angular momentum $h = v \times SP \times \sin(\alpha)$ as Newton

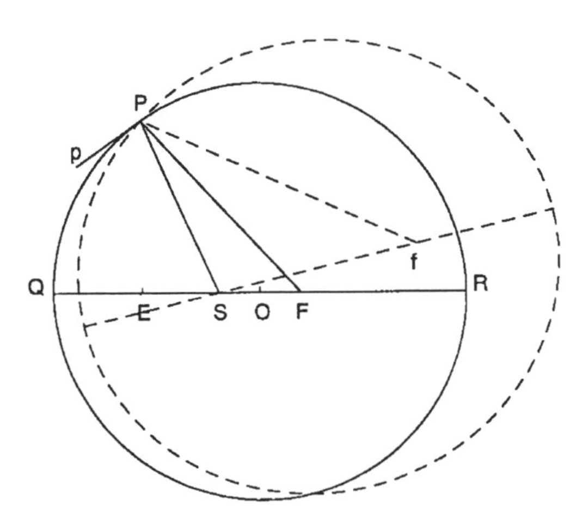

Fig. 1 Graphical illustration of Newton's Portsmouth method for the case of a single perturbation impulse. The dark lines represent the initial elliptic orbit with foci at S and F, and the dashed line the perturbed elliptical orbit after a radial impulse occurs when the revolving body is at P moving in the direction Pp.

demonstrated in Propositions 14 and 16. In order to determine the absolute value of L it is also necessary to evaluate L from a reference orbit, a point that has caused considerable confusion in the literature.[23,24]

Newton first considered the effect of a radial perturbation of magnitude V on an elliptic orbit to lowest order in the eccentricity $e = SF/QR$. Referring to Fig. 1, which is a simplified version of the diagram in the manuscript, Lemma [α] is given in the form

> "If the moon should revolve round the earth S in the elliptic orbit QPR having axis QR and foci S, F, and all the while be continuously impelled towards the earth by some force V different from its weight to the earth, but let the distance SF of its foci be indefinitely small: the motion of the apogee ensuing from that impulse will be to the mean motion of the moon round the earth in a ratio compounded of the ratio of twice the force V to the moon's average weight P and the ratio of the line SE, lying between the earth's centre and the perpendicular PE, to the distance SF of the foci."

Setting $\delta\omega$ for the change of the angle ω of the major axis of the ellipse during a small interval of time δt, and $\delta\theta$ for the corresponding change in the longitude θ of the moon, Lemma [α] implies that

$$\frac{\delta\omega}{\delta\theta} = 2\frac{V}{P} \times \frac{SE}{SF} \tag{2}$$

as Newton wrote explicitly at the conclusion of the proof of this lemma.

Newton's derivation in modern notation takes the following form. The change in radial velocity during a small interval of time δt is given by $\delta v = V\delta t$. The main effect of the perturbation is to rotate the velocity vector trough a small angle $\delta\alpha \approx \delta v/v$, where α is the angle between the tangent line Pp along the velocity and the radial line SP. Since Newton supposed that "the distance SF of its foci be indefinitely small," the orbit is nearly circular, and we have $v\delta t \approx SP\delta\theta$. Hence

$$\delta\alpha = \frac{\delta v}{v} = \frac{V\delta t}{v} = V\frac{SP}{v^2}\delta\theta\,. \tag{3}$$

Substituting the relation for the radial acceleration in a circular orbit $v^2/SP = P$, where $P = \mu/SP^2$ is the gravitational force, we obtain

$$\frac{\delta\alpha}{\delta\theta} = \frac{V}{P}\,. \tag{4}$$

The rotation of the tangent line by $\delta\alpha$ gives rise to a corresponding rotation of the line PF to the second focus of the ellipse by an amount $2\delta\alpha$ setting this focus at a new point f. This gives rise to a rotation $\delta\omega$ of the axis and a change in the distance SF between the two foci,

$$SF\delta\omega = Ff\cos(\phi') \tag{5}$$

where ϕ' is the angle between the major axis QR and PF, and $Ff = 2PF\delta\alpha$. According to Proposition 17, the distance Pf is determined by the relation (1)

$$Pf = \frac{L \times SP}{4SP\sin^2(\alpha + \delta\alpha) - L} \tag{6}$$

because L remains unchanged by the effect of a radial perturbation. For small eccentricity, $\alpha \approx \pi/2$ and the contribution of $\delta\alpha$ to this expression is second order and negligible, which implies that $Pf = PF$. Hence

$$\frac{\delta\omega}{\delta\alpha} = 2\frac{PF\cos(\phi')}{SF} \tag{7}$$

which in the limit of small eccentricity becomes

$$\frac{\delta\omega}{\delta\alpha} = \frac{2SE}{SF} \tag{8}$$

where $PE = SP\cos(\phi)$. Finally combining Eqs. (4) and (8), one obtains

$$\frac{\delta\omega}{\delta\theta} = 2\frac{V}{P} \times \frac{SE}{SF} \tag{9}$$

which is Newton's result in Lemma [α].

Setting $SE = -r\cos(\phi)$, where $r = SP$, $\phi = \theta - \omega$ is the angle of the radius vector relative to the axis of the ellipse, and $SF = 2ae$ where $a = QR/2$ is the major axis of the ellipse and e is the eccentricity, Newton's expression, Eq. (9), can be written in the form

$$\frac{d\omega}{d\theta} = -\frac{r^3}{\mu a e}V\cos(\phi)\,. \tag{10}$$

In his derivation, Newton neglected all but the lowest order term in the eccentricity e, and therefore one should set $r = a$ in this equation. However, it is straightforward to solve Newton's geometrical construction for the effect of the perturbation force V without any approximations which gives the exact result first obtained by Euler and later by Lagrange and by Laplace[5,7,26]

$$\frac{d\omega}{d\theta} = -\frac{r^2}{\mu e}V\cos(\phi) \tag{11}$$

where r is determined from the equation of an ellipse

$$r = \frac{\frac{L}{2}}{[1 + e\cos(\phi)]} . \tag{12}$$

In the Portsmouth manuscript, Newton did not evaluate the corresponding change in the eccentricity. This change can also be obtained geometrically in the same straightforward manner, and follows directly from the relation

$$\delta(SF) = Sf - SF = Ff\sin(\phi') . \tag{13}$$

Setting $SF = eQR$, neglecting $e\delta(QR)$ and substituting $Ff = 2FP\delta\alpha$, one obtains

$$\frac{\delta e}{\delta\alpha} = \sin(\phi') \tag{14}$$

and correspondingly

$$\frac{\delta e}{\delta\theta} = \frac{V}{P} \times \frac{PE}{SP} . \tag{15}$$

In similar manner, Newton considered in Lemma [β], the case that the perturbation is a transverse force W which can also be considered tangential in the limit of small orbital eccentricity. Now the change in the velocity $\delta v = W\delta t$ occurs along the direction of v, and the angular momentum changes by an amount $\delta h = SP \times \delta v = SP \times W\delta t$ while the angle α remains unchanged. Since $L = 2h^2/\mu$, we have $\delta L = (4h/\mu)SP \times W\delta t$ which, in the limit of nearly circular orbits, corresponds to $\delta L = (4/\mu)SP^3 \times W\delta\theta$. According to Proposition 17, a change in L gives a change in the magnitude of the distance to the second focus at f

$$\delta(PF) = QR \times \frac{PF}{SP} \times \frac{\delta L}{L} \tag{16}$$

and in the limit of small eccentricity, one obtains

$$\frac{\delta(PF)}{\delta\theta} = 2QR \times \frac{W}{P} . \tag{17}$$

Referring to Fig. 2, this change leads to a rotation of the major axis by an angle $\delta\omega$, where

$$SF\delta\omega = \delta(PF)\sin(\phi') \tag{18}$$

and substituting Eq. (17) in Eq. (18),

$$\frac{\delta\omega}{\delta\theta} = 2\frac{W}{P}\frac{PE}{OS} \tag{19}$$

where $OS = SF/2$, which corresponds to Newton's expression in Lemma [β].

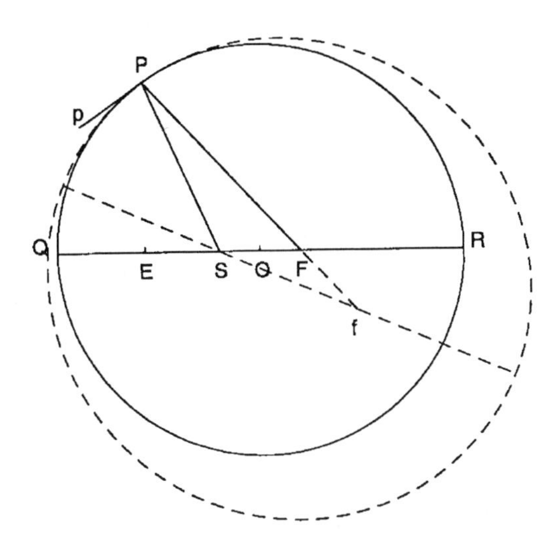

Fig. 2 Same as Fig. 1 for a tangential impulse which keeps the angle α between the radial line SP and the direction of the velocity Pp fixed. The effect of the impulse is to change the angular momentum which increases the distance PF from P to the second focus F of the ellipse.

Setting $PE = r\sin(\phi)$, $OS = ae$ and $P = \mu/r^2$, this relation corresponds to

$$\frac{d\omega}{d\theta} = 2\frac{r^3}{\mu ae} W \sin(\phi) \tag{20}$$

while the exact result obtained from Newton's construction is[5,7,26]

$$\frac{d\omega}{d\theta} = \frac{r^2}{\mu e}\frac{W \sin(\phi)}{(1 + e\cos(\phi))}\sin(\phi)[2 + e\cos(\phi)]\,. \tag{21}$$

In order to solve these equations, it is also necessary to obtain the equation for the change in the eccentricity e. From Newton's construction in Lemmas $[\alpha]$ and $[\beta]$, one readily finds[27] that to lowest order in e

$$\frac{de}{d\theta} = \frac{r^2}{\mu}[V\sin(\phi) + 2W\cos(\phi)]\,. \tag{22}$$

While there is no evidence that Newton obtained this equation, he described qualitatively the variation in the eccentricity of the orbit of the moon due to the perturbation of the sun in Proposition 66, Corollary 9, Book I of the *Principia*, and pointed out correctly that the magnitude of the variation depended on the angle $\omega - \theta'$ between the apsides of the moon and the earth-sun axis.

The variation of the angular momentum h where $h = \sqrt{\frac{\mu L}{2}}$ is obtained by integrating

$$\frac{dh}{d\theta} = \frac{r^3}{h} W \tag{23}$$

and gives rise to a variation in the magnitude of the major axis, but it was not considered by Newton in the Portsmouth manuscript.

These differential equations can be approximately solved by directly substituting for V and W the radial and tangential components of the solar perturbing force. However, this method gives rise to a puzzle whose resolution apparently evaded Newton, and may have been ultimately the real cause of his well-known remark that the lunar problem was the only problem that had made his head ache. The puzzle is the following: how could Newton's elliptic orbit solution for the moon with the earth at the center given in Book III, Proposition 28 (which corresponds to the variational solution obtained much later by L. Euler[5] and G.W. Hill[28]), emerge from a solution represented by a rotating ellipse with the earth at a focus? Newton remarked that for this special orbit "the perturbing force does not cause a rotation of the apsis,[29]" but this statement seems incompatible with his variational equations, Eqs. (9) and (19). As we shall demonstrate below, for the resolution of this paradox it is necessary to obtain also the equation for the change in the eccentricity e, an equation which is not found in the manuscripts of the Portsmouth collection.[30]

3 Hourly Motion of the Apogee

In a proposition in the Portsmouth manuscript entitled "To Ascertain the Hourly Motion of the Moon's Apogee," Newton obtained expressions for the radial and transverse components of the solar perturbation, and then substituted these in his general differential equation for the motion of the apogee. These components were provided without giving a full derivation, but the missing parts can be reconstructed from the description given in Proposition 26, Book III. The gravitational attraction of the moon to the sun is inversely proportional to the square of their relative distance $S'P$. For the purposes of his calculation, Newton expressed the magnitude of this force, equated to acceleration, as a length in the form $(S'S)^3/(S'P)^2$, where $S'S$ is the distance between the sun and the earth. Newton was careful to subtract *vectorially* from this *accelerative* force the corresponding gravitational acceleration of the earth towards the sun which in this notation has the magnitude $S'S$. Referring to Fig. 3, since the distance $S'S$ between the earth and the sun is

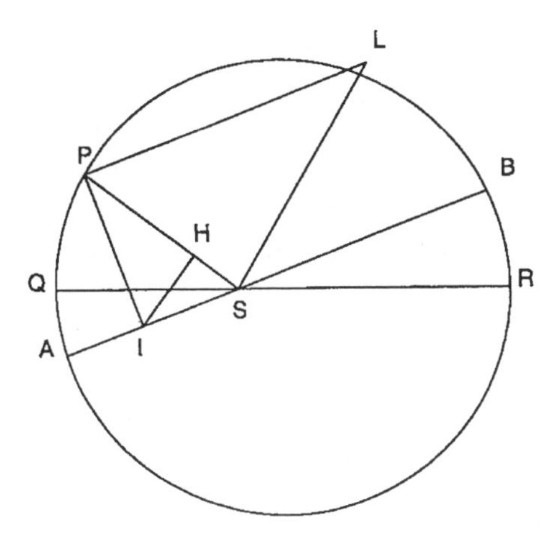

Fig. 3 Diagram illustrating the decomposition of the solar perturbation along radial and transverse directions to the earth moon axis SP. The line AB is along the earth sun axis.

much larger than the distance SP between the moon and the earth, $S'P \approx S'S - IS$, where IS is the component of SP along the earth-sun axis AB. Then $(S'S)^3/(S'P)^2 - S'S \approx 2IS$ is the component of the solar perturbation along the earth sun axis, or as Newton put it "$2IS$ shall draw the moon from the line CD" where C and D are the quadratures of the moon. Correspondingly, the component PI of SP normal to this axis gives the component of the solar perturbation in this direction. Drawing a line IH from I normal to SP, Newton then "resolves" these force components further into its radial and transverse components, which are $2HS - PH$ and $3IH$ respectively. It might be expected that at this point Newton would substitute these components for V and W in his equation for the rotation of the apsis ω, but this is not the case. Instead, he remarked that

> "These forces do not, when the moon is located in the orbit in which by their means it might revolve without eccentricity, contribute at all to the motion of the apogee."

This comment can be understood in light of a special solution for the perturbed lunar motion which Newton gave in Proposition 28 of Book III. Applying a completely different treatment of the solar perturbation based on his curvature approach, Newton found an approximate solution for the lunar orbit. This orbit is an ellipse with the earth at the *center* of the ellipse, and its major axis normal to the sun-earth axis, and the eccentricity is due entirely to the perturbation. However, he failed to distinguish the apogee of the lunar orbit from the apogee of the ellipse associated with his geometrical construction. Consequently, by not recognizing this crucial point, Newton would have been unable to derive his special lunar orbit as a solution of the Portsmouth equation.

This special solution can also be obtained from the present perturbation method, but there is no evidence that Newton attempted this derivation. Newton remarked that

> "The motion of the apogee arises from the differences between these forces and forces which in the moon's recession from that concentric orbit decrease, if they are centripetal or centrifugal in the double ratio of the (increasing) distance between the moon and the earth's center, but should they act laterally, in the same ratio triplet — as I found once I undertook the calculations."

Subsequently, Newton discussed how to implement these somewhat unclear concepts mathematically which, as we shall see, were not entirely correct. In effect, Newton realized that only terms in his equation which dependent on the intrinsic eccentricity of the orbit would contribute to the rotation of the apsis. These terms can be obtained by a systematic expansion to first order in powers of this eccentricity of the differential equation for the rotation of the apsis. However, since Newton had first calculated this equation only to lowest order in the eccentricity, it took considerable physical intuition on his part to deduce the appropriate terms. The details of Newton's procedure are discussed by Whiteside[17] in extensive footnotes to his translation of the Portsmouth manuscript, and therefore only some of the highlights will be discussed here. However, in order to examine Newton's solution and to see the source of some of its errors and shortcomings, we need to understand first the correct derivation which we will carry out here in modern notation.

Setting $r = SP$ and ψ equal to the angle PSA between the earth-moon and earth-sun axis, we have $PI = r\sin(\psi)$ and $IS = r\cos(\psi)$. Hence $PH = r\sin^2(\psi)$, $HS = r\cos^2(\psi)$, $IH = r\cos(\psi)\sin(\psi)$, and we obtain for the radial perturbation V and transverse perturbation W

$$V = g(2HS - PH) = \frac{g}{2}r(1 + 3\cos(2\psi)) \tag{24}$$

and

$$W = -g(3IH) = -3\frac{g}{2}r\sin(2\psi) \tag{25}$$

respectively. We have introduced the appropriate constant of proportionality $g = GM_S/R^3$ for the solar perturbation, where M_S is the mass of the sun and G is Newton's gravitational constant. Since the earth-sun orbit is elliptical, g varies inversely with the cube of the earth-sun distance $R = SS'$, and Newton also described in the *Principia*, the small effects of this variation on the lunar motion.

For the variation of the angle ω, we must also keep terms independent of the eccentricity e, and according to Eqs. (11) and (21)

$$\frac{d\omega}{d\theta} = \frac{r^2}{\mu e}[-V\cos(\phi) + W\sin(\phi)(2 - e\cos(\phi)] , \tag{26}$$

where $\phi = \theta - \omega$. The last term $e\cos(\phi)$ in this equation was neglected by Newton. For the variation of the eccentricity e, which was not described in the Portsmouth manuscript, we keep terms to first order in e,

$$\frac{de}{d\theta} = \frac{r^2}{\mu}[V\sin(\phi) + W(2\cos(\phi) + e(1 - \cos^2(\phi))] . \tag{27}$$

Substituting in Eqs. (26) and (27) the components of the solar perturbation force V and W given by Eqs. (24) and (25) respectively, we obtain

$$\frac{d\omega}{d\theta} = -\frac{gr^3}{2\mu e}[(1 + 3\cos(2\psi))\cos(\phi) + 3\sin(2\psi)\sin(\phi)(2 - e\cos(\phi))] , \tag{28}$$

and

$$\frac{de}{d\theta} = \frac{gr^3}{2\mu}[(1 + 3\cos(2\psi))\sin(\phi) - 3\sin(2\psi)(2\cos(\phi) + e(1 - \cos^2(\phi)))] . \tag{29}$$

To keep terms to first order in e, we approximate

$$r = r_0(1 - e_0\cos(\phi)) \tag{30}$$

and expand $e = e_0 + \Delta e + \delta e_1$, where e_0 is the unperturbed value of the eccentricity, and obtain

$$\frac{d\Delta e}{d\theta} = \frac{m^2}{4}[2\sin(\phi) - 3\sin(2\psi + \phi) - 9\sin(2\psi - \phi)] \tag{31}$$

and

$$\frac{d\delta e_1}{d\theta} = \frac{3m^2 e_0}{4}[5\sin(2\psi - 2\phi) + 4\sin(2\psi + 2\phi) - \sin(2\phi) + 5\sin(2\psi)] , \tag{32}$$

where $m^2 = gr_0^3/\mu$ is a dimensionless constant. Here $m = (T_M/T_S)$, where T_M is sidereal period of the earth-moon system and T_S is the sidereal period of the earth-sun system. Likewise, expanding $\omega = \Delta\omega_0/e_0 + \omega_1$, we obtain

$$\frac{d\Delta\omega_0}{d\theta} = \frac{m^2}{4}(-2\cos(\phi) + 3\cos(2\psi + \phi) - 9\cos(2\psi - \phi)), \tag{33}$$

and

$$\frac{d\omega_1}{d\theta} = \frac{3m^2}{4}[1+5\cos(2\psi-2\phi)+3\cos(2\psi)+\cos(2\phi)-2\cos(2\psi+2\phi)]. \tag{34}$$

The equation for ω_1 has a constant term on the right-hand-side equal to $3m^2/4$ which gives rise to a secular contribution to ω_1. Moreover, as Newton demonstrated in the Portsmouth manuscript, the appearance of a cosine term on the right-hand-side of Eq. (34) with argument $(2\psi - 2\phi)$ leads to an important correction to the magnitude of this secular contribution.

Neglecting all but the first two terms on the right-hand-side of Eq. (34), we have

$$\frac{d\omega_1}{d\theta} = \frac{3m^2}{4}[1 + \beta\cos(2\psi - 2\phi)] \tag{35}$$

where we introduced the constant $\beta = 5$. Setting $\psi = (1 - m)\theta$ and $\omega_1 = (1 - \nu)\theta + \delta\omega_1$, where $\delta\omega_1$ is assumed to be a small contribution, we have $\phi = \nu\theta - \delta\omega_1$, and expanding to first order in this quantity, we obtain

$$\delta\omega_1 = -\frac{3m^2\beta}{8(\nu - 1 + m)}\sin 2(\psi - \nu\theta) \tag{36}$$

with

$$\nu \approx 1 - \frac{3m^2}{4}\left[1 + \frac{3m^2\beta^2}{8m}\right]. \tag{37}$$

Substituting the value $\beta = 5$, one finds that

$$\delta\omega_1 = \frac{15m}{8}\sin 2(\nu - 1 + m)\theta \tag{38}$$

and

$$\nu \approx 1 - \frac{3m^2}{4}\left[1 + \frac{75m}{8}\right]. \tag{39}$$

This expression for ν corresponds to the result for the secular variation in the angle of the apsis first found by Clairaut and d'Alembert[18] (see Appendix A).

Keeping only the first term in Eq. (32) for the variation δe_1 of the eccentricity, we have

$$\frac{d\delta e_1}{d\theta} = \frac{15m^2 e_0}{4}\sin 2(\psi - \nu\theta) \tag{40}$$

and integrating this equation, one obtains

$$\delta e_1 = \gamma \cos 2(\psi - \nu\theta) \tag{41}$$

where $\gamma = 15me_0/8$.

The oscillations $\delta\omega_1$ and δe_1 correspond to the oscillations in a kinematical model introduced by Horrocks in 1641 to account for the evection, where the value of γ was obtained from observation. Since $(1-\nu)\theta$ is the mean value of the angle of the apsis ω, the argument $\nu\theta - \psi = (\nu - 1 + m)\theta$ is the angle of the sun relative to the direction of the lunar apsis, $\theta' - \omega$. Hence the period T_H of this oscillation is $T_s/2(m-1+\nu)$, where T_s is the sidereal period of the moon. Substituting the value of ν corresponding to the observed rotation of the lunar apsis which is about 3^0 per month, $1-\nu = 3/360$, and we obtain $T_H = 6.75$ months which is the period introduced by Horrocks and somewhat earlier by Kepler to account for the evection. Evaluating $(e_0 + \delta e_1)\cos(\nu\phi - \delta\omega_1)$ to first order in the small oscillation, we have

$$\delta e_1 \cos(\nu\theta) + e_0 \delta\omega_1 \sin(\nu\theta) = \gamma\cos(2\psi - \phi) \tag{42}$$

and

$$r = r_0[1 - e_0\cos(\nu\theta) - \gamma\cos(2\psi - \nu\theta)]\,. \tag{43}$$

The last term accounts for the evection which in this form has a period of $T_s/(2-\nu-2m) = 31.8$ days.

To obtain Δe and $\Delta\omega_0$ we neglect the oscillatory terms $\delta\omega_1$ and δe_1, and set $\phi = \theta - \omega = \nu\theta$ to integrate Eqs. (31) and (33), thus obtaining

$$\Delta e = \frac{m^2}{4}\left[-2\cos(\phi) + \frac{3}{(3-2m)}\cos(2\psi + \phi) + \frac{9}{(1-2m)}\cos(2\psi - \phi)\right] \tag{44}$$

and

$$\Delta\omega_0 = \frac{m^2}{4e_0}\left[-2\sin(\phi) + \frac{3}{(3-2m)}\sin(2\psi + \phi) - \frac{9}{(1-2m)}\sin(2\psi - \phi)\right]. \tag{45}$$

Hence,

$$e\cos(\phi) \approx (e_0 + \Delta e)\cos(\nu\theta) + \Delta\omega_0 \sin(\nu\theta) \tag{46}$$

and substituting Eqs. (44) and (45), we find

$$e\cos(\phi) \approx e_0\cos(\nu\theta) + \frac{m^2}{2}\left[-1 + \left(5 + \frac{28}{3}m\right)\cos(2\psi)\right]. \qquad (47)$$

To obtain the orbit in this approximation, we set

$$r \approx \frac{h^2}{\mu}(1 - e\cos(\phi)) \qquad (48)$$

where h is obtained by integrating Eq. (23), and neglecting the variation in r due to the eccentricity,

$$h = h_0\left[1 + \frac{3m^2}{4(1-m)}\cos(2\psi)\right]. \qquad (49)$$

Thus one obtains, in an expansion valid up to cubic powers in m,

$$r = r_0[1 - e_0\cos(\nu\theta) - x\cos(2\psi)] \qquad (50)$$

where $r_0 = (h_0^2/\mu)(1 + m^2/2)$, and $x = m^2(1 + (19/6)m)$.

In particular, for the limit of zero initial eccentricity, $e_0 = 0$, one recovers the special solution called the variation orbit first obtained by Newton applying his curvature method[31] in Book III, Proposition 28. This orbit was obtained later by Euler[5] and in further detail by G.W. Hill.[28]

The contributions to the evection and the variation can be combined linearly to give

$$r = r_0[1 - e_0\cos(\nu\theta) - x\cos(2\psi) - \gamma\cos(2\psi - \nu\theta)]. \qquad (51)$$

This is the "oval of another kind"[25] for the lunar orbit which Newton had been seeking, including effects of the solar perturbation to first order in the eccentricity parameter e_0, and to third order in the relative strength of the perturbation m,[32] and corresponds to the solution obtained by Clairaut and d'Alembert.[16,18]

To compare the lunar theory with observations, it is very important to realize that accurate data had been obtained on the angular position θ of the moon as a function of time t rather than on its radial distance r. This distance could be estimated only approximately from the variation of the apparent diameter of the moon. According to theory,

$$t = \int_0^\theta d\theta \frac{r^2}{h} \qquad (52)$$

and substituting Eqs. (49) and (51), we obtain

$$nt = \theta - 2e_0 \sin(\nu\theta) - \xi \sin(2\psi) - \frac{2\gamma}{(1-2m)} \sin(2\psi - \nu\theta)\,, \tag{53}$$

where $n = h_0/r_0^2 = 2\pi/\tau$, τ is the sidereal period of the moon, and $\xi = (1/(1-m))(x + 3m^2/8(1-m)) \approx (11/8)m^2 + (59/12)m^3$. The quantity nt is called the *mean anomaly* of the moon and the deviation $\theta - nt$ is the *equation of time.* The second term has an amplitude $2e_0 = 0.110 = 6^0\ 18'$, while the last term corresponds to the lunar inequality known as the *evection* first pointed out by Ptolemy, with an observed amplitude of $1^0\ 15'$.

The oscillations of the apsides and the eccentricity, Eqs. (38) and (41), are discussed only qualitatively in the *Principia*, in Proposition 66, Corollaries 8 and 9 of Book I. There, Newton gave a physical argument based on the solar perturbation force for the dependence of these oscillations on the angle $\omega - \theta'$, where θ' is the solar longitude.[34] As mentioned earlier, these oscillations were originally proposed by Horrocks in a kinematical model of the lunar orbit.[35] In the revised scholium to Proposition 35, Book III, the second and third edition of the *Principia*, Newton described this model as following from his gravitational theory, but fitted the amplitude of the oscillations from astronomical observation rather than from the theory.[36] Actually, without the equation for the variation of the eccentricity, calculated to first order in the unperturbed eccentricity e_0, Newton could not have accounted fully for Horrock's model from his gravitational theory.

Newton further argued that

> "The motion of the apogee arises from the differences between these forces and forces which in the moon's recession from that concentric orbit decrease, if they are centripetal or centrifugal, in the doubled ratio of the (increasing) distance between the moon and the earth's center, but, should they act laterally, in the same ratio tripled — as I found once I undertook the calculations."

The precise meaning of this crucial sentence is unclear. We have shown that the secular motion of the apogee is obtained from terms which are independent of the eccentricity e in Eq. (26). Thus for the radial perturbation $V(r,\psi)$, this corresponds to replacing $r^2V(r,\psi)$ in Eq. (26) by the difference $r^2V(r,\psi) - r_0^2V(r_0,\psi)$ to first order in e, where $r \approx r_0(1 - e\cos(\theta))$. Indeed, this is mathematically equivalent to Newton's own calculation. However for the transverse perturbation $W(r,\psi)$, he applied a cubic scaling factor without justification, in effect replacing $r^2W(r,\psi)$ by the difference $r^3W(r,\psi) - r_0^3W(r_0,\psi)$, which is incorrect.[37] Thus, although Newton was able to obtain the correct *analytic* form for the differential equation for ω_1, which we will call here ω_N, some of the coefficients were wrong. Written in modern notation, Newton's differential equation for ω_N takes the form[38]

$$\frac{d\omega_N}{d\theta} = \frac{3m^2}{4}[1+(11/2)\cos(2\psi-2\phi)+3\cos(2\psi)+\cos(2\phi)-(5/2)\cos(2\psi+2\phi)]\,. \tag{54}$$

Comparison with the correct equation, Eq. (34), indicates that only the coefficients multiplying the terms $\cos(2\psi - 2\phi)$ and $\cos(2\psi + 2\phi)$ are incorrect. Moreover, he correctly identified the first cosine term on the right-hand-side of Eq. (54) as the term which modifies the lowest order secular contribution to ω. Keeping only this term leads to the reduced equation for ω_N, Eq. (35), with $\beta = 11/2$. Setting $m^2 = 1/178.725$ and $\delta\theta = 13.18^0/\text{day}$ gives for the maximum daily advanced of the apogee $\delta\omega_N = 21'\ 34''$, and for the regression $\delta\omega_N = 14'\ 56''$. These values are in good agreement with the values 23′ and 16 1/3′ given in the scholium of Proposition 35, Book III in the first (1687) edition of the *Principia* quoted in our Introduction. Newton's equation can be rewritten in the form,

$$\frac{d\chi}{d\theta} = b[\cos(\chi) - n)]\,, \tag{55}$$

where $\chi = 2(\psi - \phi) \approx 2(\omega_N - m\theta)$, $b = (3/2)m^2\beta$, and $n = (1/\beta)(4/3m - 1)$. Hence

$$\frac{d\omega_N}{d\chi} = \frac{1/\beta + \cos(\chi)}{2[\cos(\chi) - n]}\,. \tag{56}$$

Newton integrated Eqs. (55) and (56) to obtain the changes $\Delta\omega$ and $\Delta\theta$ for half a cycle $\Delta\chi = \pi$, and found

$$\Delta\theta = \frac{1}{b}\int_0^\pi \frac{d\chi}{n - \cos(\chi)} = \frac{2\pi}{b\sqrt{(n^2-1)}}\,, \tag{57}$$

and

$$\Delta\omega_N = \frac{1}{2}\int_0^\pi d\chi \frac{(\frac{1}{\beta} + \cos(\chi))}{n - \cos(\chi)} = \frac{\pi}{\sqrt{(n^2-1)}}\left(\frac{1}{\beta} + n - \sqrt{n^2-1}\right). \tag{58}$$

Hence,

$$\frac{\Delta\omega_N}{\Delta\theta} = \frac{3}{4}m^2(1 + \beta(n - \sqrt{(n^2-1)}))\,. \tag{59}$$

The term $(3/4)m^2$ corresponds to Newton's perturbation result in Proposition 45, Corollary 2 (to order m^2), and the additional factor in brackets is equal to 1.924 (for Newton's numerical values $n = 3.0591$ and $\beta = 11/2$) accounting to a good approximation for the missing factor two in the earlier calculation (see quotation in the Introduction). Annually $\Delta\omega_N = 38.86^0$ is in good agreement with Newton's claim in the first edition of the *Principia* that

> "... its mean annual motion should be virtually 40^0."

Expanding the square root in powers of $1/n^2$, and substituting the expression for n, one obtains to cubic order in m^3

$$\frac{\Delta\omega}{\Delta\theta} = \frac{3}{4}m^2\left(1 + \frac{3m\beta^2}{8}\right) \tag{60}$$

which corresponds to the previous result, Eq. (37), obtained by a perturbation expansion.[39] In particular, setting $\beta = 5$ yields

$$\frac{\Delta\omega}{\Delta\theta} = \frac{3}{4}m^2\left(1 + \frac{75m}{8}\right) \tag{61}$$

which is the result, Eq. (39), obtained by Clairaut and d'Alembert[18] (see Appendix A).

4 Conclusions

The Portsmouth mathematical papers show that by 1687, Newton had developed a general method to treat the effect of perturbations on the Keplerian orbit of planets. This method corresponds to the perturbation of variation of orbital parameters developed many years later by Euler, Lagrange and Laplace.[5–8] Newton applied his method to evaluate the "motion of the apogee" of the lunar orbit to lowest order in the eccentricity. He was able to deduce, by somewhat unclear physical arguments, the correct *analytic form* for the equation describing the motion of the lunar apogee, and to solve this equation for the secular variation valid to *cubic* order in a perturbation theory in powers of m. Indeed, Newton's analytic solution for the motion of the apogee corresponds to Clairaut's and d'Alembert's solution obtained some 63 years later, after they re-discovered Newton's curvature method which he applied only to the perturbation of a special lunar orbit in the *Principia* Book III, Proposition 28. However, Newton apparently failed to realize that his equation for the motion of the apogee of the ellipse, constructed by the Portsmouth method, applied to the *physical* orbit only in the case that the unperturbed eccentricity e of this orbit is large compared to a dimensionless parameter m^2. Here, m is the reciprocal of the number of lunar cycles during a sidereal year. Thus, Newton would have been unable to understand how the special perturbed solution of the circular orbit which he obtained by the curvature method,[31] emerged from his Portsmouth method.

In his later work on the lunar inequalities, Newton apparently abandoned the further development of his remarkable Portsmouth perturbation method. However, the evidence presented here indicates that he had found a correct perturbation approach, and that apart from some technical errors he nearly succeeded in evaluating the first higher order terms in the solar perturbation to the motion of the lunar apogee.

Appendix A: The Solution of Clairaut and d'Alembert

The expression for the perturbed orbit of the moon, Eq. (51), was obtained by Clairaut and d'Alembert,[16,18] as an approximate solution for small e and m^2 of the curvature equation,

$$\left(\frac{d^2}{d\theta^2}+1\right)\frac{1}{r}=\frac{\mu}{h^2}-\frac{r^2}{h^2}\left[V-\frac{W}{r}\frac{dr}{d\theta}\right], \tag{62}$$

where

$$h^2=h_0^2+2\int d\theta r^3 W, \tag{63}$$

and V and W are the solar perturbations obtained by Newton, Eqs. (24) and (25) respectively. Substituting for r as a function of θ, the approximation of a rotating ellipse, $r=r_0(1-e\cos(\nu\theta))$ to first order in e, on the right-hand-side of Eqs. (62) and (63),

$$r^2V=\frac{m^2\mu}{2}\left[1-3e\cos(\nu\theta)+3\cos(2\psi)-\frac{9e}{2}[\cos(2\psi+\nu\theta)+\cos(2\psi-\nu\theta)]\right], \tag{64}$$

$$Wr\frac{dr}{d\theta}=\frac{3m^2\mu e}{4}[\cos(2\psi+\nu\theta)-\cos(2\psi-\nu\theta)] \tag{65}$$

and

$$h^2=h_0^2\left[1+\frac{3m^2}{2}\left[\cos(2\psi)-\frac{4}{3}e\cos(2\psi+\nu\theta)-4e\cos(2\psi-\nu\theta)\right]\right]. \tag{66}$$

This implies that, to order em^2, there are also terms proportional to $\cos(2\psi + \nu\theta)$ and to $\cos(2\psi - \nu\theta)$ on the left-hand-side of the curvature equation. Thus, to obtain these terms, an improved approximation for r must have the form

$$\frac{1}{r} = \frac{1}{r_0}[1 + e\cos(\nu\theta) + x\cos(2\psi) + \delta\cos(2\psi + \nu\theta) + \gamma\cos(2\psi - \nu\theta)] \quad (67)$$

where δ and γ are new coefficients determined by matching the corresponding terms proportional to $\cos(2\psi+\nu\theta)$ and $\cos(2\psi-\nu\theta)$, which appear in Eqs. (64)–(66). This implies that

$$\gamma = \frac{15me}{8} \quad (68)$$

and

$$\delta = -\frac{5m^2e}{8}\,. \quad (69)$$

The unexpected result is that the coefficient γ is of order me instead of order m^2e as in the case of δ. Therefore, the contribution to r of the corresponding cosine term $\cos(2\psi - \nu\theta)$, should not be neglected in the evaluation of the right-hand-side of Eq. (62). Thus, one finds that the contribution of this term gives rise to additional terms proportional to $\cos(\nu\theta)$ on the right-hand-side of Eq. (62), which therefore modified the previous evaluation of ν. In particular, r^2V gives the added term

$$-\frac{9m^2\mu\gamma}{4}\cos(\nu\theta)\,, \quad (70)$$

while $Wrdr/d\theta$ contributes

$$-\frac{3m^2\mu\gamma}{4}\cos(\nu\theta) \quad (71)$$

and h^2 contributes

$$-\frac{6m^2\gamma}{\nu}\cos(\nu\theta)\,. \quad (72)$$

Collecting all the terms which appear on the right-hand-side of Eq. (62) which are proportional to $\cos(\nu\theta)$, leads to the new relation

$$(\nu^2 - 1) = -\frac{3}{2}m^2\left(1 + \frac{5\gamma}{e}\right)\,, \quad (73)$$

and substituting Eq. (68) for γ, one obtains

$$\nu = 1 - \frac{3}{4}m^2\left(1 + \frac{75m}{8}\right) \quad (74)$$

which corresponds to Eq. (39).

Notes and References

1. *A Catalogue of the Portsmouth Collection of the Books and Papers Written By or Belonging to Sir Isaac Newton*, the scientific portion of which has been presented by the Earl of Portsmouth to the University of Cambridge. Drawn up by the syndicate appointed the 6 November 1872. (Cambridge University Press, 1888). The members of the syndicate were J.C. Adams, G.D. Liveing, H.R. Luard and G.G. Stoke. The content of Newton's manuscript with an English translation from the original Latin and detailed commentaries by D.T. Whiteside appears in *The Mathematical Papers of Isaac Newton* (ed.) D.T. Whiteside (Cambridge University Press, 1974), 1684–1691, Vol. 6, pp. 508–538.
2. It is of considerable interest that in the 1684–1685 revised treaty of *De Motu*, Corollaries 3 and 4 were not included in Proposition 17 (see *The Mathematical Papers of Isaac Newton* (ed.) D.T. Whiteside (Cambridge University Press, 1974), 1684–1691, Vol. 6, p. 161, but these appeared in the 1687 edition of the *Principia.* In his book *Introduction to Newton's Principia* (Cambridge University Press, 1971), I.B. Cohen concluded that "there is no way of telling whether they (Newton's Portsmouth manuscripts) date from just before or just after publication of the *Principia*" (p. 120). However, the outline of Newton's Portsmouth perturbation method in these corollaries clearly indicates that Newton must have developed this method sometime between 1685 and 1687. Whiteside attributed the Portsmouth manuscripts to late 1686.
3. D.T. Whiteside, "Newton's lunar theory: From high hope to disenchantment," *Vistas in Astronomy* **19** (1976): 317–328.
4. Isaac Newton's *Philosophiae Naturalis and Principia Mathematica*, the third edition with variant readings (eds.) Alexander Koyré and I. Bernard Cohen with the assistance of Anne Whitmann (Harvard University Press, 1972), Vol. 2, pp. 658–661.
5. Leonhardi Euleri, *Opera Omnia Series Secunda, Opera Mechanica et Astronomica* (eds.) L. Courvoisier and J.O. Fleckenstein (Basileae MCMLXIX), Vol. 23, pp. 286–289. Starting with the equations of motion, written as second order differential equations in polar coordinates, Euler assumed that the solution for the orbit is described by an ellipse with time-varying orbital parameters p, e and ω, where p is the semilatus rectum of the ellipse, e is the eccentricity, and ω is the angle of the major axis. Then he obtained first order differential equations for e and ω by imposing two

constraints: that $p = h^2/\mu$, where h is the angular momentum, and that $E = \mu(e^2 - 1)/2p$, where E is the time-varying Kepler energy of the orbit. In modern notation, $\mu = GM$ where M is the sum of the mass of the earth, and the moon and G is Newton's gravitational constant. It can be readily shown that Euler's constraints lead to the same definition of the ellipse described geometrically by Newton in the Portsmouth manuscript. I would like to thank C. Wilson for calling my attention to Euler's work.

6. Curtis Wilson, "The work of Lagrange in celestial mechanics," in *The General History of Astronomy, Planetary Astronomy From the Renaissance to the Rise of Astrophysics* (eds.) R. Taton and C. Wilson (Cambridge University Press, 1995) 2B, pp. 108–130.
7. P.S. Laplace, "A treatise of celestial mechanics" translated from the French by Henry H. Harte (Dublin, 1822), pp. 357–390. Laplace obtained the differential equations for the time dependence of the orbital parameters by evaluating the time derivative of the vector $\vec{f} = \vec{v} \times \vec{h} - \mu\vec{r}/r$, where $\vec{f}$ is a vector along the major axis of the ellipse with magnitude $f = \mu e$. The construction of this vector was first given in geometrical form by Newton in Book I, Proposition 17, and in analytic form by Hermann and Johann Bernoulli in the *Mémoires de l' Acad'emie Royale des Sciences*, 1710. Laplace's derivation of the variation of orbital parameter is in effect the analytic equivalent of Newton's geometrical approach in the Portsmouth manuscript.
8. P.S. Laplace, "Sur la théorie lunaire de Newton," *Mécanique Céleste* (Paris, Imprimiere Royale MDCCCXLVI), Tome V, Chapitre II, p. 438.
9. The equivalence of Newton's Portsmouth method to the variation of orbital parameters method of Euler, Lagrange and Laplace has not been generally appreciated in the past. For example, in "Newton's lunar theory: from high hope to disenchantment," *Vistas in Astronomy* **19** (1976): 321, D.T Whiteside wrote that "Newton, it is clear from his private papers (the Portsmouth manuscripts) passed in some despair to introduce the additional Horrocksian assumption — one only approximately justifiable from the three-body dynamical problem whose more accurate solution he thereby ceased to control, then and ever after — ... The truth as I have tried to sketch it here is that his loosely approximate and but shadowly justified way of deriving those inequalities which he did deduce was a retrogressive step back to an earlier kinematic tradition which he had once hoped to transcend ..." On the contrary, Newton's geometrical

construction is not an approximate but an exact implementation of the effects of a general perturbation force on Keplerian motion. Its analytic evaluation leads to the first order differential equations for the variation of orbital parameters later formulated by Euler, Lagrange and Laplace. In *Newton's Principia for the Common Reader* (Oxford University Press, 1995), p. 252, S. Chandrasekhar guessed that Newton must have known these equations, and in *The Motion of the Moon* (Adm. Hilger, Bristol and Philadelphia, 1988), p. 57, A. Cook wrote that "there are clear indications in Newton's unpublished treatment of the effects of the Sun upon the Moon that he was aware of relations equivalent to Lagrange's planetary relations." In fact, as shown in Sec. 2 in the Portsmouth manuscripts, one finds the derivation of the differential equation for one of these parameters, the angle of apsis ω, which is evaluated correctly to lowest order in the eccentricity e. Moreover, the solution of this equation for the case of the solar perturbation on the lunar orbit is nearly correct.

10. Craig B. Waff, "Isaac Newton, the motion of the lunar apogee, and the establishment of the inverse square law," *Vistas in Astronomy* **20** (1976): 99–103; "Newton and the motion of the moon: an essay review," *Centaurus* **21** (1977): 64–75; "Clairaut and the motion of the lunar apse: the inverse-square law undergoes a test," in *Planetary Astronomy From the Renaissance to the Rise of Astrophysics, The General History of Astronomy* (eds.) Rene Taton and Curtis Wilson (Cambridge University Press, 1995), 2B, pp. 35–46.
11. Philip P. Chandler, "The *Principia's* theory of the motion of the lunar apse," *Historia Mathematica* **4** (1977): 405–410.
12. Curtis Wilson, "The Newtonian achievement in astronomy," in *The General History of Astronomy. Planetary Astronomy From the Renaissance to the Rise of Astrophysics* (eds.) R. Taton and C. Wilson (Cambridge University Press, 1989), 2A, pp. 262–267.
13. S. Chandrasekhar, *Newton's Principia for the Common Reader* (Clarendon Press, Oxford, 1995). For my review of this book, see *American Journal of Physics* **14** (1996): 450–454.
14. Laplace, Ref. 8, p. 419.
15. Curtis Wilson, "Newton on the moon's variation and apsidal motion: the need for a newer 'new analysis'," in *Isaac's Newton's Natural Philosophy* (eds.) J. Buchwald and I.B. Cohen (Cambridge MIT Press, forthcoming).

I am indebted to Professor Wilson for sending me a copy of this article during the completion of my work.

16. F. Tisserand, "Théorie de la lune de newton," *Traité de Mécanique Céleste* (Gauthier-Villars et Fils, Paris, 1894) Tome III, Chapitre III, pp. 27–45.
17. *The Mathematical Papers of Isaac Newton* (ed.) D.T. Whiteside (Cambridge University Press), 1684–1691, Vol. 6, pp. 508–535.
18. Curtis Wilson, "The problem of perturbation analytically treated: Euler, Clairaut, d'Alembert," in *Planetary Astronomy From the Renaissance to the Rise of Astrophysics, The General History of Astronomy* (eds.) Rene Taton and Curtis Wilson (Cambridge University Press, 1995) 2B, p. 103
19. In fact, Proposition 17 corresponds, in geometrical form, to the Laplace vector $\vec{e} = (\frac{1}{\mu})\vec{v} \times \vec{h} - \vec{r}/r$, where $\vec{r}$ is the position, $\vec{v}$ is the velocity and $\vec{h}$ is the angular momentum vector (see Ref. 7).
20. Isaac Newton, *Mathematical Principles of Natural Philosophy*, 3rd Ed. translated by I.B. Cohen and Anne Whitman (Los Angeles, Berkeley, London, University of California Press, 1997).
21. In the 1690's, Newton considered a revision of his demonstration of Proposition 17 which would enable him to derive the effect of a perturbation normal and tangential to the velocity more readily and without restrictions to small eccentricities. He then included and extended version of Corollaries 3 and 4 explaining in detail with a new diagram how to evaluate the effect of the perturbation. He also added Corollaries 5 and 6 giving the results of his derivation for the rotation of the apsis in the case of a tangential and a normal perturbative impulse (see Ref. 17, pp. 559–563). Newton's derivation can be reconstructed and it can be seen that his results are not quite correct.
22. In the scholium to Proposition 35, Newton attributed to Horrocks only the model of a simple elliptical orbit for the motion of the moon around the earth, and to Halley the addition of an epicycle to account for the variation of the eccentricity and the rotation of the apsis. Actually, it was Horrocks and not Halley who introduce this epicycle to account for the *evection*, and inequality of the moon discovered by Ptolemy.
23. J. Herivel, *The Background to Newton's Principia* (Oxford University Press, 1965), p. 291, footnote 32, Herivel commented that "... this [auxiliary orbit] represents a weakness of the present method ..."
24. Whiteside, Ref. 17, p. 278, footnote 5, Whiteside remarked that "... an auxiliary orbit is not ... at all necessary."

25. Whiteside, Ref. 17, p. 519.
26. J.M.A. Danby, *Fundamentals of Celestial Mechanics*, 2nd. Ed. (Willmann-Bell, Richmond VA, 1989), p. 327. Alan Cook, *The Motion of the Moon* (Adm, Hilger, Bristol and Philadelphia, 1988), p. 57.
27. The exact result is

$$\frac{de}{d\theta} = \frac{r^2}{\mu}\left[V\sin(\phi) + \frac{W}{(1+e\cos(\phi))}[2\cos(\phi) + e(1+\cos^2\phi)\right]. \qquad (75)$$

28. *Collected Mathematical Papers of G.W. Hill* (Carnegie Institute of Washington, 1905), Vol. 1, pp. 284–335.
29. Whiteside, Ref. 17, p. 517.
30. In Lemmas [α] and [β], Newton referred to "the motion of the apogee ensuing from that impulse" as if it were the apogee of the moon's orbit, which is not strictly correct. Even modern textbooks of celestial mechanics do not make the important distinction between the apogee of the physical orbit and the apogee of the ellipse in the variation of orbital parameters method.
31. M. Nauenberg, "Newton's early computational method for dynamics," *Archive for History of Exact Sciences* **46** (1994): 221–252.
32. The relation between time t and longitude θ obtained from Eq. (51) takes the form

$$\frac{t}{\tau} = \theta - 2e_0\sin(\nu\theta) - \xi\sin(2\psi) - \frac{2\gamma}{(1-2m)}\sin(2\psi - \nu\theta) \qquad (76)$$

where $\tau = 2\pi r_0^2/h_0$. The last term corresponds to the lunar inequality known as the *evection* first pointed out by Ptolemy.
33. Newton's remark can be interpreted as setting $r = r_0$ on the right-hand-side of Eq. (26) for ω, which leads to Eq. (45) for $\Delta\omega_0$ with $\omega = \Delta\omega_0/e_0$. This gives a purely oscillatory contribution to ω, and together with the corresponding contribution Δe, which Newton apparently did not evaluate, this gives rise to the special lunar orbit as can be seen by taking the limit $e_0 = 0$ in Eq. (50).
34. In Proposition 66, Corollary 9, Newton concluded that

> "... therefore in the passage of the apsis from the quadratures ($\omega - \theta' = \pm\pi/2$) to the syzygies ($\omega - \theta' = 0, \pi$) it is continually augmented, and increases the eccentricity of the ellipse; and in the passage form the syzygies to the quadrature it is continually decreasing, and diminishes the eccentricity."

35. Curtis Wilson, "predictive astronomy in the century after Kepler," in *The General History of Astronomy, Planetary Astronomy From the Renaissance to the Rise of Astrophysics* (eds.) R. Taton and C. Wilson (Cambridge University Press, 1989), 2A, pp. 197–201. Horrocks' model for the lunar orbit can be written, for small eccentricity e_0, in the analytic form $r = r_0[1 - (e_0 + \delta e)\cos(\nu\theta - \delta\omega)]$, where $\delta e = \delta e_0 \cos(\xi)$, and $\delta\omega = (\delta e_0/e_0)\sin(\xi)$. Here, $\xi = 2(\omega - \phi') \approx 2(\psi - \nu\theta)$ where $\phi' \approx m\theta$ is the longitude of the sun. This model is justified by Newton's gravitational theory and gives an important contribution to the lunar inequalities due to the solar perturbation. According to the perturbation theory $\delta e_0/e_0 = 15m/8$.
36. In the revised scholium to Proposition 35, Book III, Newton stated that

> "By the same theory of gravity, the moon's apogee goes forwards at the greatest rate when it is either in conjunction with or in opposition to the sun, but in its quadratures with the sun it goes backwards; and the eccentricity comes, in the former case to its greatest quantity; in the latter to its least by Cors. 7, 8 and 9, Prop. 66, Bk. I. And those inequalities by the Corollaries we have named, are very great, and generate the principle which I call the semiannual equation of the apogee; and this semiannual equation in its greatest quantity comes to about $12^0 18'$, as nearly as I could determine from the *phenomena* [italics mine]. Our countryman, Horrocks, was the first who advanced the theory of the moon's ..."

In Corollaries 7 and 8, Proposition 66, Newton gave a qualitative explanation for this motion of the moon's apogee due to the perturbation of the sun, claiming it was based on results given in Book I, Proposition 45, Corollary 1. However, these results were obtained for the case of radial forces only, and are therefore strictly not applicable to the solar perturbation which is not a purely radial force with respect to the earth as a center, and which depends also on the angle ψ. According to the differential equation for the motion of the lunar apogee which appears in the Portsmouth manuscript, Eq. (35), this rate depends on the relative angle between the moon's apogee ω and the longitude θ' of the sun, where $\omega - \theta' = \psi - \phi$. It reaches a maximum value when $\omega - \theta' = n\pi$ where n is an integer, and a minimum when n is an odd integer divided by 2, in accordance with Corollary 8. In Corollary 9, Newton gave a qualitative argument for the variability of the eccentricity, corresponding to Eq. (41) (and not Eq. (44) as claimed by Chandrasehkar, Ref. 13, p. 251), but there is no evidence that he obtained this quantitative result from his "theory of gravity" (according to his theory, the maximum variability of the apogee is $15m/8 = 8^0 2'$ instead of $12^0 18'$ as quoted in the scholium to Proposition 35). Although the Horrocksian model was probably the inspiration for his Portsmouth method, in the end Newton was able to account partially for this model from his dynamical principles.

37. For the contribution of the radial component of the perturbation force to the secular motion of the apogee (Whiteside, Ref. 17, p. 521), Newton took the difference $(1-(QO/SP)^3)(2HS-PH) \approx (3x/SP)(PH-2HS)$, where $x = SP - QO$. In our notation $SP = r = r_0(1 - e\cos(\phi))$, $QO = r_0$, $x = -r_0 e\cos(\phi)$ and $2HS - PH = (r/2)(1 - 3\cos(2\psi))$. However, for the corresponding contribution of the transverse component of the force Newton assumed the difference $(1-(QO/SP)^4)3IH \approx (12x/SP)IH$, where $IH = (r/2)\sin(2\psi)$. The introduction here of a quartic rather than cubic power in the scale factor QO/SP is not correct. Newton considered also other powers which has led to the accusation that he was "fudging" his equation to get agreement with observation (see Whiteside Ref. 17, p. 518, footnote 26. However, we would like to stress Newton's insight, that the motion of the apogee is determined by the difference of properly scaled perturbation forces acting on unperturbed circular and elliptic orbits, is essentially correct.
38. In Newton's notation, this equation appears as the ratio of

$$6HS \times SE^2 - 3PH \times SE^2 + 24IH \times PE \times SE \tag{77}$$

to $178.725P^2$ for "the total motion of the apogee arising from both forces to the moon's mean motion." Here $HS = r\cos^2(\psi)$, $SE = r\cos(\phi)$, $PH = r\sin^2(\psi)$, $IH = r\cos(\psi)\sin(\psi)$, and $P^2 = 1/r^2$ (Whiteside, Ref. 17, p. 521)
39. In Ref. 13, Chandrasekhar claimed that the exact solution of Newton's equation for the mean motion of the apogee, either with Newton's parameter $\beta = 11/2$ or the correct parameter $\beta = 5$, "comes nowhere near resolving the discrepancy." However, this statement is based on an erroneous result given in Ref. 13, Eq. (36), p. 453.

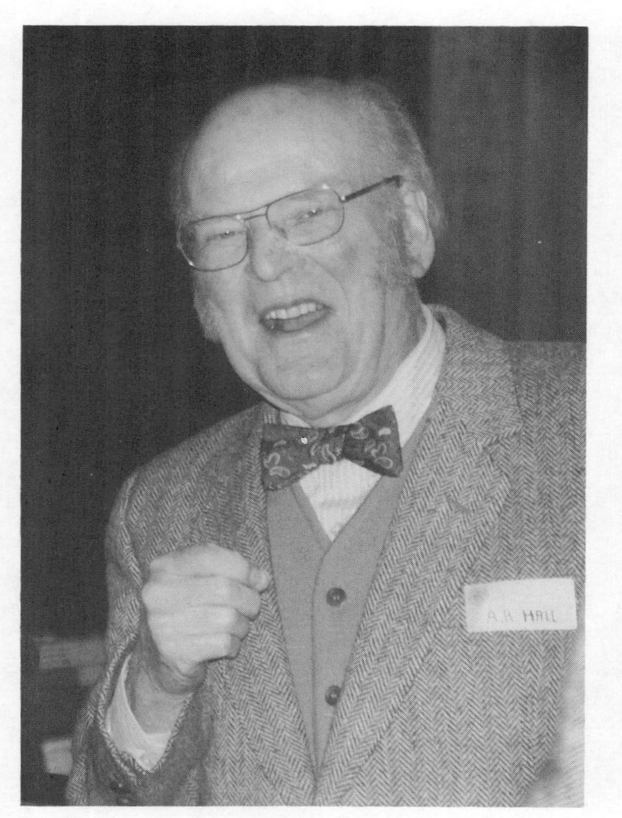

R. Hall

M. (Boas) Hall

Review and Reminiscences

A. RUPERT HALL
14 Ball Lane, Tackley OX5 3AG
Oxford, U.K.

Friends and Colleagues!

I have been invited to open the commentary on today's Newtonian proceedings. I believe it is just about ten years since the last symposium devoted to Isaac Newton was held here at the Royal Society, on the occasion of the *Principia*'s tercentenary. Some of today's speakers spoke then also. I wish there could be an event of the same character *every* ten years, or better still, every five. The example of the Harriot Seminar shows that such an object is not unattainable, and there are certainly sufficient scholars active in Newtonian studies to maintain such a series. It is important that the momentum gained by our studies during the last half-century should not be lost, and that it should not be supposed that all has been done, that it is profitable to do. The contrary is the case.

I do not mean solemnly to review all our day's work, only to make a point or two about it. It is tempting on an occasion of this kind to use it to present an *apologia pro vita sua*, but I shall endeavor not to fall into this vein. Really, the two most important words I can utter this afternoon are *thank you*: I give my thanks first of all to the organizers of this pleasant and profitable day, who have been laboring at their task for long months, and finally brought it to this profitable fruition; and I reiterate thanks to the various speakers, who have brought many accessions to our knowledge in a short space of time.

Next, a word or two about absent friends and fallen comrades, mentioning first Professor Curtis Wilson. It is a great pity that although we shall have his contribution to this symposium, ill health has prevented his being here. There are others, not all of whose names have been mentioned today, to whom I am now happy to pay public tribute on so fitting an occasion as this. First, I beg to recall a great mentor of mine in matters historical, and especially intellectual history, Herbert Butterfield of Cambridge (1900–1979), who opened to many of us new prospects for the historiography of science as the history of thought about nature, as Bernard Cohen was quick to recognize long ago.[1] To him I am indebted not only for providing a model to emulate but also for benevolent influences on the course of my life. Next to recall another of the same time and of even greater impact on our field of studies: I mean Alexandre Koyré (1892–1964), a scholar especially noted (as Butterfield of course was not) for Newtonian researches. Koyré was both a deep scholar and an acute philosopher; none in our field have used this combination of talents to greater effect.[2] His own idealist vision of the history of science was powerfully impressed upon a whole generation. I personally profited much from his encouragement and criticism. After the publication of our book on the *Unpublished Scientific Papers of Isaac Newton* (1962) he wrote to us, I remember, regretting that he had spent much labor upon transcribing *De gravitatione & aequipondio fluidorum* from the manuscript when we were about to put it into print.

Coming nearer to our own time, a word must be said of two American students of Newton's writings only recently lost to us, Sam [Richard S.] Westfall and Betty Jo Teeter Dobbs, the former the major exponent of Newton's biography in our age — I well remember his consulting us in Los Angeles nearly forty years ago on whether this might be a suitable theme for a life's work — the latter a major exponent of Newton's alchemical papers. Newton's chemistry or alchemy — call it what you will — was very far from being an exact or mathematical science (though it bears customary evidences of his attention to quantitative measurement) and accordingly we have heard nothing of it today. Both Dobbs and Westfall sought for a unifying consistency in Newton's thinking about the natural world, in which — somehow — there was room for both the whimsicality of alchemy and the exactitude of geometry: if we (as historians) accept 99% of the *Principia* and 90% of *Opticks* as constituting segments of the bedrock of modern physical science, while setting aside the rest and with it a fair slice of Newton's surviving manuscript legacy, as extravaganza outside science, then we are (on such a view as theirs) creating a highly tendentious, partial and unhistorical image of Newton's intellectual strivings and achievements. It would be absurd to link a proposition in the *Principia* with the hunt for volatile metals, but both were Newton's concern and there are steps of transition between them. We have to thank these two scholars, as well as others living and dead, for profound and stimulating work on these manuscript materials, which are far from easy to interpret, and for at least beginning to create a holistic picture of Newton bringing in sources too long neglected. I think I may have been the first person actually to handle Newton's chemical notebook (C. U. L. MS. Add 3975) since the end of last century; by coincidence I happened to send for it (nearly fifty years ago!) very soon after it had been restored to its place in the Cambridge University Library from the University Registry archives, where it had lain forgotten. H.R. Luard, the University Registrar in the 1880s, had been one of the Syndicate which (in effect) brought Lord Portsmouth's Newton papers to the University (where the mathematical and scientific portions of them were to remain) and prepared the printed catalog of them; hence, I suppose, the misplacement may have occurred.[3]

Before I go on, let me add the name of yet a third scholar not long lost to us, who almost uniquely penetrated (in recent years) into other dark corners of Newton's literary output: I mean Frank Manuel.[4] Now, fortunately, we have a number of young scholars like Scott Mandelbrote working on Newton's theology, though Newton as a historian seems not to have attracted successors to Manuel.

The point I would make, by mentioning on this occasion not only scholars who have re-examined Newton's place, his supreme place, in the history of mathematics and physical science but also those who have delved into his far more obscure and tortuous non-scientific investigations, which seem to have contributed little or nothing to our twentieth century world, is that both aspects of the Newtonian legacy are now being studied by younger scholars. Although the non-mathematical and non-experimental writings of Newton (both published and manuscript) were little explored during the rapid and successful burgeoning of Newtonian studies during the middle and post-middle years of this century, a broader, more philosophical attitude is now evident. It is of course of the highest importance, and is the highest tribute that we can pay to Newton's intellectual stature, to try to solve the problem that *Principia* and *Opticks* set before us, and to read these works in the context of an evolutionary linear development of (say) celestial mechanics, fluid mechanics or color theory, but there remain always other problems to tease us, problems that arise as soon as we view Newton's intellectual life as a whole in all its vast variety, and relate it to the contemporary context.

I for one do not wish to see the rise of two or more divergent schools of Newtonian scholarship, one group of scholars confining themselves to the *Prophecies of Daniel and John* or "Diana's Doves", the other group not looking outside *Principia* and *Opticks*, and perhaps each group claiming, as tends to happen in these cases, that the Newton with whose writings they are concerned, is the "true" Newton. That is why I agree strongly with Bernard Cohen in deploring that too well-known dictum of Lord Keynes about the last of the magicians;[5]

it encouraged the discovery of a "new" or "true" Newton who was clearly not the author of *Principia* and *Opticks* — for who could think of these as books of magic? — but who was clearly credulous, obsessively religious and as deeply enmeshed in outmoded scholarly fantasies as George Eliot's Casaubon. I do not think we improve our comprehension of the Newton who was unquestionably a principal founder of our modern intellectual world by pairing him with a second personality who was desperately striving to perpetuate the old world. Such an approach rather retards than promotes a holistic image of Newton.

Newtonian scholarship should stand, it seems to me, appropriately in its Cambridge context, upon a stable tripod. One of its legs, very robust and strong, is the leg created by mathematical historians. With this body of knowledge we have largely been concerned today, more particularly with Newton as a mathematical physicist. The second leg is formed by the history of Newton's experimental researches, which of course are represented (in print) in *Principia* as well as in *Opticks* and the chemical papers. I need hardly emphasise to this audience the strong integration effected by Newton between experiment and mathematics. Today, we heard Alan Shapiro's exposition of another foray into the Newtonian archives, this time in search of fresh light upon his published accounts of optical diffraction. As he has shown on many occasions, Alan has a fine ability to reconstruct from Newton's laconic notes on experiments the course and incidents of the line of inquiry that Newton followed; how he achieved successes (which sometimes proved delusive!), how he sometimes met rebuffs and had to try a new cast, how the numbers and (sometimes) the rather odd statements that appear in the printed texts were arrived at in the first place, how Newton took care about testing the quantitative predictions derived from his expectations about the way things are with the results obtained from actual measurement. I can only concur with the sentiment expressed this morning: we all eagerly await the second volume of Newton's optical papers.

Taken into the private laboratory, as it were, under Alan's guidance one meets a Newton who seems a good deal less dogmatic and confident than was the author of *Opticks*, a book in which he had to face the public. And I suspect this to be true of other of his experimental researches in optics. This still uncertain Newton, seeking his way through a problem, who was exploring rather than legislating, choosing between multiple possibilities both mathematical and experimental that presented themselves to him, had to foresee and test his path leading towards the results he hoped to achieve. We have indeed seen today how evolving experiments and evolving mathematical analysis sometimes forced Newton to change his mind about the probable nature of things. We have seen that such uncertain, exploratory gropings into the unknown compelled him to try one mathematical approach, then reject it in favor of a second or third or to test an experimental model only to find it wanting, again resorting to a second or third attempt.

Sometimes, Newton's concision of argument made important ideas obscure as with Mike Nauenberg's reference to *Principia*, Book I, Proposition 17, Corollaries 3 and 4, proposing a successful mathematical route that Newton might have followed at one point in the exploratory stages of composing the *Principia* (1686?, 1687?), but for whatever reason did not actually adopt. Just as with Alan Shapiro's investigation of the experimental work behind *Opticks*, so also here, it is most illuminating that recent studies have begun to penetrate beneath the polished, adamantine surface of Newton's great printed works to the foundations of his public formulations, to reveal the possible alternative arguments or positions that Newton considered and rejected, for example. Bruce Brackenridge's recent work (including his paper today) has similarly reconstructed mathematical procedures employed by Newton of which only traces now remain. In the past, scholars — myself in a very minor way among them — have concerned themselves with the historical evolution and context of Newton's scientific texts, their various methods of proof, and with the development of texts in their successive later editions.[6] We have also had a fair amount of work, sometimes benefiting from posterity's hindsight, on paralogisms and conceptual weaknesses in Newton's writings, and naturally a good deal of critical analysis of the metaphysical foundations of Newtonian natural philosophy. Scholars move forward rather like a flock of sheep across a wide field; having digested long ago problems about Newton's obligations to his predecessors, we are now looking closely at the details of the structural evolution of the texts themselves, from their earliest rudiments as they can be discovered; we now know that neither *Principia* nor *Opticks* sprang like Minerva from the head of Jove: they are a palimpsest of investigation and tentative endeavors.

I cannot resist one further glance at *Opticks*, because Alan's paper today decisively clarified an old problem. The unfinished state of that work, emphasized by the Queries that Newton tacked on to the first edition text, has long been a puzzle. Newton's excuse for abandoning Book III, Part I (as it now is) — that he lacked leisure to pursue further experiments — has long seemed a lame one, especially since we knew that Newton had set his optical work aside years before he quitted Cambridge, and before he stopped chemical experimentation. (And we no longer suppose him to have been mentally incapable of continuing his researches into diffraction, had he so chosen.) What Alan has clearly demonstrated is that he was intellectually blocked in groping towards a theory of diffraction by the fixed set of his mind; he could not create and therefore could not use the theoretical tools required for understanding the paradoxical phenomena of diffraction. With Alan, we have seen how Newton, on the basis of an imperfect empirical picture, at first sought for a relatively simple "force" solution; experiment following the failure of this approach simply led him into insuperable difficulties. *Opticks* brings its readers, very honestly, to the limits of Newtonian physics; it is hardly a matter of careless omission that there is no word of either "fits" or diffraction in Robert Smith's *Opticks* (1738). I must quote now a sentence from Alan's paper that was not read this morning:

> "Newton's unpublished papers on diffraction provide an unusually detailed look into the way he carried out an experimental investigation and utilized his data in calculations to deduce and reject laws."

The character of historical inquiry defined here seems to me to have typified our day's work: we have been given glimpses — more is hardly possible — into the way in which Newton created his sciences and expressed his results in his printed works. As in previous studies by today's speakers, we have today learned more about the processes of Newton's thought in mathematical and experimental science, without which knowledge we can scarcely hope for a deep knowledge of the printed books themselves.

Discussion

In answer to a question about my own introduction to Newtonian studies:

For reasons which I need not go into, as an absolute ignoramus eager to learn, in those far-off unprofessional days when all things seemed open to a serious inquirer, I decided to write a thesis on the history of ballistics in the seventeenth century, though I was ill-enough equipped for the more advanced mathematical aspects of such a task.[7] I was fairly soon deep into the well-known treatment of projectile motion by Galileo, and the less tractable discussions of the more complex ideas of such mathematicians as Viscount Brouncker, Christiaan Huygens, James Gregory and Isaac Newton. I also looked into the rather unsatisfactory literature that then existed on science and early warfare, and indeed on science and technology in general. Why, I wondered, had a number of first-rate mathematicians addressed themselves to the special problems of ballistics treated as a problem in fluid mechanics, while practical men or men of lesser capacities, wrote about the art of shooting with great guns? I came across the paper by the Russian Hessen on "The social and economic roots of Newton's *Principia*" (1931),[8] offering the solution that "the bourgeoisie" was conscious of gunnery and navigation (among other things) as defective branches of technical knowledge, so Newton as a mathematician choosing problems to tackle naturally selected those that society presented to him. Not being quite satisfied with this solution — for after all Newton was almost notoriously uninterested in celestial navigation and never saw the sea (so unlike Huygens!) — I went, naively you may judge, to the University Library (where I knew there was a Newton archive) to seek for any light this might throw on his choice of topics for treatment in the *Principia*; naively, I say, because although I had already looked through the printed correspondence of Christiaan Huygens and other relevant material in the *Oeuvres Complètes*, I had very little knowledge of the information given by scientists, or not given, about their choice of problems. Apart from correspondence, Newton in fact left negligible material other than drafts connected with the first edition of the *Principia.* The correspondence does a little illuminate his interest in ballistics. Necessarily, he was ignorant of Huygens's private researches, otherwise (it is clear) he was well-informed about the existing discussion of this topic in print. More importantly, I was fascinated by Newton's notebooks, somewhat drily summarized in the Portsmouth catalog, for in them one could see something of Newton's reading and thought as a very young man, and notes of his earliest experiments in optics. Thus it was that I came to prepare my first paper,[9] published in *The Cambridge Historical Journal* by Herbert Butterfield in 1948. No one before that time seems to have considered the possibility of a systematic investigation of Newton's early intellectual development, in which field much has since been done by Westfall in *Never at Rest* (1980) and by McGuire and Tamny.[10]

A question was also put about the examination of my thesis in 1950. My first thesis director was Sir George Clark (1890–1979), then Regius Professor of History at Cambridge. He wrote a useful little volume, critical of the Hessen thesis, *Science and Social Welfare in the Age of Newton* (1937), and long before that had introduced an excellent piece on science into his general survey, *The Seventeenth Century* (1929, rev. 1947), a book described by Clark's official obituarist as "still the best single-volume history available."[11] Much later, he was to prepare the history of the College of Physicians. Sir

George was transferred back to Oxford as Provost of Oriel College in 1947; I was then placed under Frank Puryer White, a mathematician and librarian of St. John's who had formed a notable collection of early mathematical books. The University appointed Sir George Clark as one of my examiners, together with Dr. Charles Singer whom I had not previously met. I am perhaps unique in having been examined in the Athenaeum Club down the road from here, regarded as a convenient central meeting point for all concerned. Singer, whose lungs were weak (he put this down to years of travel on the old *steam* District Railway in London, on his way to and from the City of London School), had long lived at Kilmarth — subsequently the home of a far more celebrated writer, Daphne du Maurier — near Par in Cornwall, where I was often his guest in later years. For he evidently approved of me, and had a most benevolent influence upon my career; we remained colleagues and friends to the end of his life. I would not claim, however, despite the eminence of these two historians, that the manner of examination of my thesis — later published as *Ballistics in the Seventeenth Century* (1952) — in any way reached the high levels of professionalism that prevails today.

Since I have recorded indebtedness to a number of scholars who had made notable contributions to the historiography of science, let me add just one more name, that of Charles E. Raven, Master of my College, Christ's, whose biography of John Ray will long stand as a monument.[12] Raven first proposed that the College should encourage my venturing into the history of science in 1946 and I had the advantage of attending some meetings he organized a little later, attended by a number of senior members of the university. If I add an acquaintance with Joseph Needham that developed rather later — in the immediate post-war years he was with UNESCO in Paris — I think I have recalled the names of all the very considerable scholars whose kindness and concern was so valuable to me in the five years before I took my Ph.D.

All of these, and other younger scholars with whom I had some contact, worked on the technical history of science, though it would be absurd to suppose that such authors as Needham and Raven, different as their writings are, wholly ignored the intellectual and social contexts of natural history and natural philosophy. If I may in conclusion state a *credo* it is essential for the technical historiography of science to continue: an historiography written as far as may be appropriate in the language of the sciences themselves. Contrary to a tradition that goes back at least as far as Auguste Comte, the history of science is not susceptible of any particular philosophical interpretation, whether of the working scientist's or of the historian's adoption. We all recognize the fact that at times scientific research and exposition *has* been influenced by particular philosophies, positivism for example, and equally the historiography of science has been influenced in a similar way. But however temporarily stimulating such slanted renderings of the history of science may be, they are products of fashion. The solid history of science, by whatever route we may hope to reach it, must be a structure independent of fashions in philosophy. Still less, it seems to me, can the nature and consequences of man's exploration of the universe through the millennia be treated as a mere social artefact, conditioned and directed by the character and structure of different societies. Again, we are all aware that a great deal of the scientific research carried out during the last couple of generations has been "goal-orientated"; society trains people for this role, it pays them for filling it, state or other social institutions (corporations, trusts) largely determine what kinds of laboratories are built and what kinds of problems are addressed within them. I dispute none of this. I would only add that if such a modern structure of science is a social machine, it is nevertheless also a community; a highly professional, international community of which those who administer are members as well as those who experiment and teach. In principle, a man or a woman can start as a lab. assistant and end as a Chief Scientific Adviser, just as (in the old Royal Navy) every Admiral had once been a midshipman. But the more important objection to the social machine of science as a universal historical model is its novelty: it had no existence before this century. The science of the past, though for the last three hundred years or so the product of a community of scientists, was highly individualistic; for this reason our forerunners as historians of science tended to make history the story of individuals and of institutions that functioned as corporate individuals. Nor would I deny that the individuals who framed science lived in a context which exerted its own influences upon them: these we may wish to balance against the effect upon a person's work of tradition and professional definition. New research is often an outgrowth from accepted knowledge. Not to grow tedious: of course, we all agree that history is not just the story of great men, but neither is it just an analysis of social pressures and contextual conditioning. It is therefore essential that the technical history of science should continue to flourish, as we have seen it flourish in the papers presented today, continuing a tradition of scholarship begun by our predecessors (whose achievements we should by no means set aside as of little value or wholly irrelevant to modern circumstances but rather seek to refine and extend). Of course, it is a worthy facet of historiography to reconstruct so far as we can, the context and conditions of scientists in the past, to study the resources available to them and the limitations to their enquiries, but it is only in the technical history of science — which is, in my own view, a department of intellectual history — that we approximate as closely as is possible to the minds of past scientists and natural philosophers, so that we can comprehend as exactly as may be their problems in their own terms and the processes by which they sought to resolve them.

Notes and References

1. In a review of *The Origins of Modern Science* (London, 1949) in *Isis* **41** (1950): 231–233: "Mr. Butterfield has rendered historians of science a very special service in showing them what their field may eventually become."
2. On Koyré, see the bibliography in I. Bernard Cohen and René Taton (eds.) *Mélanges Alexandre Koyré* (Paris, 1964), I; his lecture on "Newton and Descartes" was published in his *Newtonian Studies* (Cambridge, Massachusetts, 1965), pp. 53–200.
3. H.R. Luard, G.G. Stokes, J.C. Adams and G.D. Liveing, *A Catalogue of the Portsmouth Collection of... Sir Isaac Newton* (Cambridge, 1888).
4. Frank E. Manuel, *Isaac Newton, Historian* (Cambridge, Massachusetts, 1963); *idem, A Portrait of Isaac Newton* (Cambridge, Massachusetts, 1968); *idem, The Religion of Isaac Newton* (Oxford University Press, 1974).
5. John Maynard Keynes, "Newton the Man," in *Royal Society Newton Tercentenary Celebrations* (Cambridge, 1947), p. 27.
6. A. Koyré, "Pour une édition critique des oeuvres de Newton," *Révue d'Histoire des Sciences* **8** (1995): 19–47; A. Rupert Hall, "Correcting the *Principia*" *Osiris* **13** (1958): 291–326; Alexandre Koyre and I. Bernard Cohen, *Isaac Newton's Philosophiae Naturalis Principia Mathematica: The Third Edition with Variant Readings* (Cambridge, 1972).
7. Obvious (though limited) secondary works were P. Charbonnier, *Essais sur l'histoire de la Balistique* (Paris, 1928) and H.J. Tallqvist, *Oversikt av Ballistikens Historia* (Helsingfors, 1931).
8. Published in Anon, *Science at the Crossroads* (London, 1931).
9. A.R. Hall, "Sir Isaac Newton's Notebook, 1661–1665," *Cambridge Historical Journal* **9** (1948): 239–250.

10. J.E. McGuire and M. Tamny, *Certain Philosophical Questions: Newton's Trinity Notebook* (Cambridge, 1983).
11. Geoffrey Parker in *Proceedings of the British Academy*, Vol. 66, p. 414 (1980).
12. C.E. Raven, *John Ray* (Cambridge, 1950); also *idem, English Naturalists from Neckham to Ray* (Cambridge, 1947); *Natural Religion and Christian Theology* (Cambridge, 1953). I may add here also that my historical education was enriched by an invitation to attend the seminars of Professor M.M. Postan.

Derek Thomas Whiteside

How Does One Come to Edit Newton's Mathematics?*

DEREK THOMAS WHITESIDE

Department of Pure Mathematics, Cambridge University
16 Mill Lane, Cambridge CB2 1SB, U.K.

Professor Whiteside (henceforth DTW) began his informal talk with his National Service[1] during 1954–1956 in the Fifth Tank Regiment, then based at Barce in Libya on the northern fringe of the Sahara Desert. Amid the often mind-blowing heat of shade temperatures up to 160°F, from late 1955 the question of what he was going to do loomed ever more hugely. Then a good school friend Peter Hall,[2] wrote suggesting that he apply to pursue post-graduate studies at Cambridge, in his own St. Catharine's College. DTW thereupon filled in the application form and sent it off. In February 1956, he had still no reply when a broken left collar bone, irredeemably shattered, returned him to England for a bone graft on it. Three months later, now burdened with a rigid heavy plaster cast which encased his torso and left arm and held his left forearm rigidly before his face, while he was spending extended sick leave with another school friend in Somerset, he went up to Cambridge to find out what had happened. DTW sketched in his appearance at this time, not least the dirty old outsize raincoat donated by a friendly farmer from a scarecrow in one of his fields, which he pulled over his cast to protect it. When a porter showed him into the Senior Tutor of "Cat's," Tom Henn,[3] the latter peered dimly through pebble-glasses, and rapped out "Take your coat off, man!" DTW's response of "I'm afraid I can't, sir" was his entry to Cambridge. He learned long after, that his original application had been rejected, the letter of rejection having failed to follow him from Libya to his English address.

But what to study? Away in Libya he had put down "French Seventeenth Century Philosophy" on his original application form, but three days after his

*Owing to a long drawn out family tragedy and his own serious ill health since the symposium, it has not been possible for Professor Tom Whiteside to provide a manuscript of his address for inclusion in this volume. What is given here has been put together by an editor [RHD] on the basis of a tape recording made at the symposium, supplemented by a number of discussions with him by telephone, and has been accepted by him. The same applies to the "Notes and References" given at the end of this paper.

arrival in Cambridge to start his first academic research year, he decided abruptly against this choice. What instead? He remembered that, as a schoolboy, he had written a prize-winning essay on the "The History of Mathematics," so on the spur of the moment he switched to research on that topic and Professor R.B. Braithwaite[4] of King's accepted to supervise him. Since the latter was Knightbridge Professor of Moral Philosophy, DTW was registered for his three academic years (1956–1959) as a research student in the Faculty of Moral Science,[5] with thesis topic: "Some Aspects of the Growth of Mathematical Ideas of Space and Time in Seventeenth and Eighteenth Century England." Braithwaite's supervision was not well-matched to DTW's needs and the next academic year 1957–1958, he was switched to receive "supervision" from Dr. Michael A. Hoskin,[6] then a research fellow at Jesus College — a very pleasant arrangement. As DTW also told us, "He (Hoskin) was very nice to me, not so much as a technical supervisor but as a great friend, encouraging me when I was down in the dumps, which most of the time I, as every other lonely Cambridge research student at that time, was." The *Cambridge University Reporter* noted this change of supervisor in 1957, along with a change of thesis title to be: "Currents of mathematical thought in England in the late seventeenth century, with special reference to concepts of number, space and time." For 1958–1959, this supervision was mainly by correspondence, since Hoskin had taken up a lectureship in the History of Science at Leicester University in 1958.

When DTW set to work he soon found that he could place no trust in the existing secondary histories of mathematics; those in English were poorly researched or tailored their findings to a preferred overall viewpoint, while those others, most notably Cantor's *Geschichte*, written in German, French and Italian were already badly dated. So he started at rock bottom, first using good editions of individual mathematicians — really these were only those of Descartes, Fermat and Huygens — as were available. To fill in the complicated texture of mathematical development in the seventeenth century he was put *in lieu* over the two-and-a-half years from late 1956 to skimming through the content of some 400 to 500 original works, the majority to be discarded, but the remainder then studied in fine detail, the results of that research being gathered in folder after folder. He found it remarkable that many of the copies of the books he studied had most of their pages uncut, so that nobody could possibly have read their substance even though some were such well-known treatises as Euler's two volumes on calculus. Originals which he could not find in the University Library he read in the awesomely near-complete Wren Library at Trinity College, with a mere dozen exceptions which he had to go to the Bodleian at Oxford to study. Came Easter 1959 and the government scholarship which did not near cover even his expenses at Cambridge (he did hard manual work at Christmas and Easter, and slaved seven days a week for three months on Blackpool "Prom" during the summers of 1957 and 1958, often till well after dark, to make up the difference) was to terminate in June. A Leverhulme Fellowship afforded a life line to continue his researches from September on, but by quarterly cheques to be paid in arrears. Desperation! From his folders between 2nd and 30th May, he raced out a first version of his thesis in 16 chapters, and then left on the 31st, owing his bank 31 shillings 9 pence, to amass money to cover the coming September to November during which he would be back in Cambridge, a more than usually indigent scholar. He wrote up his thesis over several weekends in that autumn, smoothing out its rough verbal edges (and cutting out the three final chapters, which seemed to him insufficiently researched) and submitted it, typed up, early in December. The published version[7] of this exploration of "Patterns of Mathematical Thought in the Later Seventeenth Century" differs from the thesis only by a limited number of minor corrections made in galley proof.

DTW went on briefly to outline how he was first introduced to the hoard of Newton's scientific papers in the Cambridge University Library (CUL). He had already skim-read through Jones' 1711 edition of several of Newton's mathematical papers, and not been especially impressed. One morning early in May 1958, while talking to Mr. Pink (then Western Manuscripts librarian, whose kingdom was the Anderson Room where at that time you could read both rare books and manuscripts, and indeed anything held in CUL's "Special" collections from Penguin paperbacks to — with the university librarian's explicit permission — pornography), DTW offhandedly asked "I suppose you haven't any papers of Newton's here?" Mr. Pink looked at him for a moment, went out and came staggering back with a huge armful of the cherry-red boxes in which CUL's holdings of them were then kept, maybe eight altogether and banged them down before him with something like "Is that enough to be going on with?" to which DTW in all pomposity of youth answered they might take him a couple of weeks. The extracts from those and further boxes, which a year later he included in his thesis, were an eye-opener to the rest of the world.

In October the same year DTW received a letter written in a hand worse than any manuscript he had ever seen, so far illegible that he could not even decipher the writer's name. It was, fortunately, on C.E.G.B.[8] notepaper giving its London address (at Blackfriars) and a phone number, when he rang this and explained he had just received a letter which he could not read. He was put through to a secretary who said "I keep telling him not to write his own letters when nobody can read them!" but quickly arranged that DTW, still wholly in the dark, should come along at 3.30 a few days later. When he duly arrived on cue bearing his piece of illegible paper, a huge ex-Guardsman, all of six feet six inches, barred his way, looked briefly at the letter and said he knew nothing about it, but when pressed marched off to check with the receptionist inside, a young girl of five feet nothing. The glass doors cut off all sound within, and it was as in a silent film that DTW out on the pavement saw the massive man wilt under the tongue-lashing which the girl gave him. The ex-Guardsman then came out again meekly to ask DTW, in all the finery of a dirty outsize duffle coat, to please go in. The two inside walls, which from the outside had seemed made of sheet metal, housed (he now saw) metal doors. The girl, having telephoned someone, unlocked one of them, ushered him inside, pressed a button labelled "PENTHOUSE," then locked him in ... Instant panic assailed him, till he realized it was a lift. But going up to a penthouse? — such things were not the lot of ordinary people like him!

When the lift doors opened on the top floor, DTW was welcomed by what he described as a fashion-plate model, probably the most beautiful woman he had ever seen. Except that she was extremely cross. "I'm always telling Harold not to do this! It's *highly* inconvenient for you to come today!" Then she briskly told him that he was allowed ten minutes with "Harold," after which she would bring tea in on a tray, and then he had to go.

On entering, DTW saw a wizened old man, framed in silhouette against a large vertical skylight behind, who with great difficulty got up from the desk at which he was sitting, reached out to shake his hand with a "Name's Harold Hartley. What's yours?" In some confusion, and still not knowing who "Hartley" was, he stumbled out "Tom Whiteside," to receive an impatient "Know too many Toms already. What's the 'D' for?" "Derek, sir." And hence it came that Hartley was the only man DTW has ever willingly let call him by his first name. And of course Hartley said that DTW in return should always call him "Harold." Not that DTW had still any idea of who he was.

It quickly became clear to DTW that Hartley had come to grill him, but this in his inimitable way, and with many a zany moment. Even his first question, an apparently mundane "Where were you born?" with its prosaic answer "Blackpool, Lancashire" was given the astonishing riposte — this from one apparently a retired high-up with the CEGB — "Oh! I built the station there when I was with the L.M.S.!" There followed other questions to which DTW no longer remembers his replies. And of course after ten minutes the secretary came in, purposively rattling tea-cups and looking at DTW, to be shooed away — and this several times more when she tried repeating the ploy, with a desperate plea at the last that she had had to put "Downing Street" on hold — with an "I know your tricks, woman! It's really interesting talking to this man!" And on and on went the conversation between the two until almost 6 p.m in the evening.

As Hartley[9] began to tell something of his own life, he opened up a whole new world to DTW: one populated by people of power, both public and private, and enriched by anecdote. After taking a first degree at Oxford, he had gone on to become a tutor in chemistry at his College, Balliol.[10] On the coming of the First War he found, then in his mid-30s, that he was rejected as too old when he tried to volunteer at the Oxford recruiting office. But his father was good friends with the Colonel of a Cavalry Regiment. Hartley had, unfortunately, never learned to mount and sit a horse but he made the seven-day deadline set by the colonel to do so after his own individual fashion. After the Regiment was sent to the Western Front in November 1914, Second Lieutenant Hartley was under an almost daily barrage until the next February when the Germans started pumping mustard gas into the British trenches. The word was then passed around the B.E.F. (to officers only and not to "other ranks," of course) that anyone who knew anything of its properties, and how to neutralize these, should come forward. Hartley proved to be the only one who knew anything about "gas", and he was ordered to collect a team together to combat this brand new military weapon: a team which subsequently designed the gas mask used by the British and other armed services in both World Wars. But protocol demanded that no lowly lieutenant should address officers of staff rank, let alone Field Marshal Haig and his generals. So it was between February and October 1915 that Hartley received promotion after promotion till he attained the permanent rank of Brigadier General, in charge of Chemical Warfare: "The fastest rise to that rank on merit in the whole history of the British army" he proudly stated.[11]

DTW already had more than an inkling that this old man he chatted to in that C.E.G.B. penthouse room was one who had achieved greatness, largely unsung as it might be. He ought to have known of Hartley's efforts to build up the British chemical industry in the 1920s and later;[12] of his passing to be Chief Consulting Engineer to the L.M.S.[13] in the early 1930s[14] (during which time he less than grandly rebuilt both of Blackpool's stations, as he had told DTW earlier); of, foreseeing the future, becoming head of the joint Air Service set up by the four big railway companies, this soon to be Imperial Airways joining the whole Empire with its flying boats; of, after the outbreak of the Second War, chairing the amalgamated National Railways; of passing in 1946 to head the reformed B.O.A.C., initially using converted Lancasters; and the like of B.E.A. when it broke away from its parent shortly afterwards in 1946, returning to head B.O.A.C. for 1947–1949 and fighting for it to have the best aircraft then available (American), making it highly competitive with the world's aircraft companies. One might begin to ask what did this man not do? DTW certainly should have remembered that a profile of Hartley had appeared in the second issue, just appeared, of *New Scientist*.[15] During that afternoon of their first meeting, indeed, Hartley began subtly to turn the conversation towards popularized science, remarking for instance how he had had to teach "Prince Philip" some science, especially new trends in its development. Then he mentioned some lectures he had given on the history of chemistry[16] which he had "long ago" — in 1901 in fact! — contracted with Oxford University Press to write up, but which he had never started to revise for publication. Then suddenly, as it at once became clear what was the true reason why Hartley had invited him along, he thrust over the table to DTW several bunches of paper and said "What do you think of these? I'm not at all happy with them."

"These", Hartley at once explained, were contributions to a book of essays on prominent founding fathers of the Royal Society,[16] which were due to be published by it in 1960 as part of its tercentenary celebrations. Hartley had been given the job of editing this book, *The Royal Society: Origins and Founders* as its title had been agreed to be. All this to a 26-year-old who had only two years of postgraduate research behind him! Hartley said that most of the contributions he had received were not up to standard, remarking that "Scientists have no sense of history," and he asked DTW, "Would you look at them, and tell me what you think? This man Summerson: all he writes about is Wren's architecture. And Scott's piece on Wallis has so little substance. The one on the first President, Brouncker, is the worst of the lot! Could you help me rewrite them? 'Fraid I cannot give you any credit for doing it. But if you would like to contribute some short article of your own, that would be O.K."

So it was that after more than two hours' conversation with a man who during that time had apparently kept the Prime Minister dangling at the end of a phone, DTW bore back with him to Cambridge a wad of handwritten and typewritten drafts for articles in a coming Royal Society Commemoration volume of which he had never before heard. He tried his best to criticize constructively and make helpful suggestions as and when he saw need. Scott's article on Wallis was poor, and the unknown contributor who wrote about Brouncker had stretched thinness and repetition far beyond acceptable limits — only afterwards did Hartley tell him that it was he himself who had cobbled it together. Virtually all the improvements which DTW suggested in rewrite and revamping were accepted by their authors in the versions published. Sir John Summerson was an internationally recognized authority on Wren's architecture, much famed for his prose style. To paper over the fact that he said almost nothing about Wren's science, what could be done? Hartley persuaded DTW to insert an article of his own on "Wren the Mathematician," effectively his first publication[16] (The unattributed bibliographies which appear after the "Wallis" and "Brouncker" articles are also his).

DTW went on to say that he had always found Hartley a thoroughly honest and utterly fascinating man and stressed his deep debt to him. Until about 1970 (he died in 1972), Hartley continued to pull strings for him, behind the scenes, many of which he himself knew nothing about, save for their outcome.[17] He told us that he never again went without money, though never with much money, during the rest of Hartley's life.

DTW closed his remarks by saying that there were of course others who had been of vital importance for the success of the task he had undertaken, to edit the mathematical papers of Newton — Herbert Turnbull[18] (the first editor in a real sense of the Newton *Correspondence*), Michael Hoskin[6] (who was DTW's supervisor during 1957–1959 and then lecturer in History of Science at Cambridge from 1959), and above all Adolf Prag (the Senior Mathematics Master at Westminster School), who privately supported him through the crucial early years of inner self-doubt which all research scholars have to weather through. To these he had given their due elsewhere, in the title pages, dedications and prefaces of the eight volumes of the Newton *Mathematical Papers*, and he had no time to repeat those here. But none had the lightning flash generated in his first encounter with Hartley, that first meeting so nearly prevented by an ex-Guardsman at the C.E.G.B. door.

A final word of praise came from DTW to Rupert Hall for summarizing so well their joint opinion of the high quality of papers presented at the meeting (even though he time and again felt the urge to look over his shoulder and see who this man "Whiteside" was, whom so many spoke about in such glowing terms), and a quick call for a round of applause for Bruce Brackenridge and Michael Nauenberg, each of whom had worked so hard over so many months to bring the meeting to be, and then he was done.

These remarks generated some interesting discussion.

A voice in the audience asked about Newton's Laws of Motion. DTW replied that Newton originally drafted six laws, but dropped one of them almost at once. Two others were combined to make one law, being concerned with angular motion, and Newton realized only at a rather late stage that, with central forces, this could be deduced from the Three Laws of Motion which we know today. DTW then said that the Third Law is separate from the First and Second Laws; the idea that action and reaction are of equal magnitude but with opposite directions was already familiar to scientists very early in the seventeenth century, for example in optics for the hypothetical mechanism of eyesight visualized as a particle of light knocking out a particle of the retina and in the everyday experience of games involving the collisions of balls. He said that, in his view, the First and Second Laws really go together; one defines inertial straight motion and the other is concerned with deviations from it. Furthermore, that Hooke, through having an intuitive notion of the Second Law already in 1660,[19] had a good deal of priority over Newton in this. But Newton did not really need Hooke, DTW said; already in 1665, Newton had achieved an exact understanding of the measurement of centrifugal force by v^2/R. However,[20] Newton's discovery of angular momentum conservation came in the winter 1679/1680, after Hooke had communicated to him his not yet mathematical thoughts on orbital dynamics. Newton told Halley in 1686[21] that it was after Hooke's correspondence with him that he had discovered the origin of Kepler's area law.

Somebody else asked DTW for his critique of the Laws of Motion. He replied as follows:

"What do you want? I could give you the *history* of them. As I just said, Laws One and Three are not really Newton's. There were also forerunners to Newton, like Marcus Marci, a Czech scientist who wrote on Mechanics in the 1640s but whom few people of his own day, let alone ours, had ever heard of, and François d'Aiguillon, a French scientist who wrote on Optics in 1617 — Alan Shapiro knows more than I do about these earlier seventeenth century histories — but we really have no idea at all about whether or not Newton read, or even knew of, these earlier writings."

He went on to say that the Second Law is of the greatest importance. There Newton did something which Hooke did not know how to do: to propound a unified theory of the motions of the planets and their moons about the Sun under gravitational inverse square law forces.[22] Although Hooke had a muddy appreciation of the use of the Second Law in planetary dynamics, he had only a sketchy idea of how to give mathematical bite to the notion. DTW said that he had always felt that Hooke tried to cloak this crucial deficiency in his way of reasoning by setting up about and around it a smokescreen of words.

Notes and References

1. DTW was born at Blackpool on 23 July 1932. His secondary education was at Blackpool Grammar School, whence he went to Bristol University in October 1951, receiving a starred First Class Honours B.A. degree in July 1954, specializing in Latin and French as his major subjects but taking also (this after a giant struggle on his part) a little mathematics and philosophy.
2. Peter Geoffrey Hall, born on 19 March 1932, was in DTW's class at Blackpool Grammar, but went on to St. Catharine's College, Cambridge, where he specialized in Geography and Geology, taking his B.A. degree in 1953 and his Ph.D. in 1956. He is now Director of UCL's School of Public Policy and was knighted in June 1998 for his contributions to town and country planning.
3. Tom (Thomas Rice) Henn was the Senior Tutor of St. Catharine's College, Cambridge, for 1956 to 1957. His passion was for Irish poets and poetry, present and past, and he was a close friend of W.B. Yeats.
4. Richard Bevan Braithwaite, born 15 January 1900, was Knightbridge Professor of Moral Philosophy in the University of Cambridge from 1953 up to his retirement in 1967. He had an active interest in the mathematical theory of games, and the fundamentals of mathematics and of scientific explanation, as well as in philosophy.
5. "Moral Science" is the term used in Cambridge University for what is "Philosophy" in other Universities.
6. Michael (Anthony) Hoskin received a B.A. degree in pure mathematics from the University of London and then went to Peterhouse, Cambridge, in 1952, after his National Service at the Aldermaston Laboratory, to do research on algebraic geometry, receiving his Ph.D. degree in 1956. He held a research fellowship at Peterhouse for 1956/1957 and at Jesus College,

Cambridge, for 1957/1958 and took up a lectureship in the History of Science at Leicester University in 1958. He returned to Cambridge in 1959, as lecturer in the History of Science. He was a fellow and tutor at St. Edmunds, Cambridge, from 1965 to 1969, and was then elected fellow in the History of Science at Churchill College, where, since 1988, he is now an emeritus fellow.

7. "Patterns of mathematical thought in the later seventeenth century," in *Archive for History of Exact Sciences* **1** (1961): 179–384. This paper has eleven chapters, with a foreword and a bibliography.
8. The Central Electricity Generating Board (C.E.G.B.) came into being on 1 April 1958. After retirement from B.O.A.C. in 1949, Hartley became Chairman of the Electricity Supply Research Council of the British Electricity Authority until 1952, remaining as its Deputy Chairman until 1954. In 1961, he was appointed a Consultant to the C.E.G.B. for the rest of his life.
9. An excellent, succinct biography of Sir Harold (Brewer) Hartley, by one of his former research students E.J. Bowen, may be found in the Dictionary of National Biography 1971–1980, pp. 387–389. A longer biography giving more details about his scientific work and his impact on Science Policy in Britain has been published by A.G. Ogston, another former research student of his, in *Biographical Memoirs of Fellows of the Royal Society*, Vol. **19** (1973), 348–373. See also his entry in "*Who Was Who*" for 1971–1980.
10. Harold Brewer Hartley entered Balliol College, Oxford, in 1897, completing his B.A. degree in Chemistry in 1900. When the Balliol tutor (Sir John Conroy) died in December 1900, Hartley was elected in his place in May 1901. Also the University appointed him the first Duke of Bedford's Lecturer in Physical Chemistry, a topic not well represented at Oxford previously. Hartley was a great believer of the importance of history for a proper understanding of any subject. In the Lent term of 1901, he gave a special series of lectures entitled "Studies in the History of Chemistry," and then signed a contract with Oxford University Press for their publication in book form, as he remarked in its preface. However, he was so busy with immediate problems that its final manuscript was not delivered to Oxford University Press for many years. It was published in 1971, so he saw it in print before he died (on 9 September 1972).

11. Ogston[9] has noted that the statement "This officer possesses, to an outstanding degree, the ability to impress officers of senior rank," appears in Hartley's Army record.
12. As a result of his First War experience in charge of Chemical Warfare for the Third Army and of his thorough inspection, just after the War, of the chemical side of German wartime industries, Hartley saw chemical engineering as the most important undeveloped area in British industry and joined the Society of Chemical Engineers in 1922. He also joined the Board of the Gaslight and Coke Company, remaining a member of it until 1945 and establishing recognition of the role of science in gas-making. University graduates were also brought in to develop the efficient use of gas and coke and the improvements they achieved galvanized the whole industry. He also took an interest in the Fuel Research Board of the Government Department of Scientific and Industrial Research, joining the Board in 1929 and becoming its Chairman from 1932 to 1947. He stimulated a survey of national fuel resources, carried out jointly with the Geological Survey and colliery owners and, later, with the National Coal Board.
13. L.M.S. (London, Midlands and Scottish Railway Company), one of the three railway giants in Britain before their nationalization in 1948, was a merger of the London and North West Company and the London and Midlands Company, together with two smaller rail companies.
14. This involved him resigning from his tutorial fellowship at Balliol in 1931, in favor of a research fellowship with membership of its governing body, which he held until 1941, when he was elected an honorary fellow.
15. Profile: Sir Harold Hartley, *The New Scientist* **1** (2), 29 November 1956, pp. 31–32.
16. *The Royal Society: Its Origins and Founders*, (ed.) Sir Harold Hartley, F.R.S. (The Royal Society, London, 1960), pp. 275, ix. D.T. Whiteside's paper on *Wren the Mathematician* is to be found on pp. 107–111.
17. The acknowledged sources of financial support for the *Mathematical Papers of Isaac Newton* project are: the U.K. government (through a research student-ship for 1956–1959 and a research fellowship for 1961–1963, from its Department of Scientific and Industrial Research), the Leverhulme Trust (fellowship for 1959–1961), and tenure employment by Cambridge University from 1963 (made possible in part by contributions from the Sloan Foundation and the Master and Fellows of Trinity College). DTW was University Reader in the History of Mathematics in 1976–1987, and

has been University Professor of the History of Mathematics and Exact Sciences since 1987.

18. Herbert Westren Turnbull, born 13 August 1885 to a Yorkshire family, his father being an H.M. Inspector of Schools. Regius Professor of Mathematics at the University of St. Andrews, 1921–1950. F.R.S. 1932. Retired to Millom, Cumberland, in 1950, moving from there to Grasmere in 1959 where he died on 5 May 1961 and is buried.
19. Hooke presented his ideas concerning the Second Law at a Royal Society meeting in May 1666, publishing them in 1671 in a small book entitled *An Attempt to Prove the Motion of the Earth.* He communicated them to Newton by letter in November and December 1679.
20. M. Nauenberg, Newton's Early Computational Method for Dynamics, *Archive for History of Exact Sciences* **46** (1994): 221–251.
21. *The Correspondence of Isaac Newton*, Vol. 2, 1676–1687. (ed.) H.W. Turnbull (Cambridge University Press, 1960), pp. 434–441.
22. Isaac Newton, *Philosophiae Naturalis Principia Mathematica*, Book III.

Curtis Wilson

From Kepler to Newton: Telling the Tale

CURTIS WILSON
St. John's College, Annapolis, Maryland, U.S.A.

In the midst of intent scholars like yourselves, the teacher in me quakes. Scholarship and teaching are always threatening to split asunder and go their separate ways.Teachers fall back on stories; stories can mislead; well-worn stories are often quite false. High up behind us looms the figure of the scholar, raising an admonitory finger, reminding us how recalcitrant history is to neat packaging in words. The editors of the *Correspondence* and the *Mathematical Papers* have set the high standards that the rest of us must aspire to meet.

I first dipped into Vols. 1 and 2 of the Newton *Correspondence* in 1966, while engaged in a study of Kepler's *Astronomia Nova.* A sentence in a letter from Newton to Halley of June 1686 brought me up sharp: "Kepler knew the Orb to be not circular but oval & guest it to be elliptical."[1] The time-worn story had it that Kepler established *three* empirical laws, and these laws were somehow the basis of Newton's system.[2] The word "law", a transplant from theology, was used sparingly by Newton. He later applied it to Kepler's third rule, but never to the elliptical orbit.

My aim in 1966 was to find out how the case stood with Kepler's first two rules, posing as empirical truths. I had been teaching at a "great books college" for 18 years, and attempting with students to read Ptolemy, Copernicus, Kepler and Newton (a Quixotic enterprise, no doubt). Our Kepler was Book IV of the *Epitome of Copernican Astronomy,* which has much to say about Kepler's three famous rules. As a Whiggish simplifier, I wanted it to tell me how Kepler discovered them; instead, it gives a causal account that violates elementary physics, and is painfully *ad hoc.* Even after reading Koestler, and Koyré, I remained dissatisfied. Koyré claimed that the orbit could be determined by distance-determinations *exactement.*[3] But surely there was *some* observational error? Evidently, I needed to read the *Astronomia Nova.*

At St. John's (the "great books college") I might never have gotten round to it. But a family circumstance took me to California, where I wangled a non-tenure appointment at UCSD. There I was made to understand that I had to hump myself: perish as we all must, but for heaven's sake, publish! So, in the summer of 1966, I set out to read and fathom — if I could — the first

60 chapters of Kepler's *Astronomia Nova*, with the further aim of getting an article accepted in a few months. For guidance, I had Max Caspar's German translation, with introduction and notes. Caspar said of Kepler's first two laws that they were "aus der Erfahrung bewiesen,"[4] proven from experience. Once more, that vague claim!

Let me not, at this point, lose you in the dark inner penetralia of the *Astronomia Nova.* Kepler's persistent effort to accommodate empirical data, as subsumed, in particular, in his hypothesis vicaria, is to be praised; equally to be acknowledged is the theory-laden-ness of his war on Mars. Somewhere in the midst of that war, I found, Kepler had learnt that his final ellipse *plus* the area rule would yield acceptable equations of center. But he had not yet *arrived* at the ellipse. How he did so must be inferred from Chaps. 58 and 59, where surface intricacies can befuddle. He showed that the solar distances of the planet in the apsides and at the quadrants fitted the ellipse. His further claim to have excluded alternatives to the ellipse[5] at other angles from the Sun is invalid, as Tom Whiteside has shown in Ref. 6. The empirical grounding for the leap to the ellipse was soft.

Not irrelevant to our judgment of Kepler's ellipse are questions about his causal explanations. His notion of inertia was pre-Galilean: a body moves only if pushed; something issuing from the Sun pushes the planets about. To obtain an elliptical orbit, he had to postulate a simple harmonic oscillation along the radius vector. As the cause of this oscillation, he hypothesized magnetic fibers in the body of the planet, such that the planet was alternately attracted and repelled by the Sun. The Sun itself had to be a sort of magnet, with its surface as one pole. The magnetic fibers in the planet had to maintain a nearly constant orientation at right angles to the line of apsides. Thus, for example, the Earth's interior would not participate in the eastward rotation of the Earth's crust. Jeremiah Horrocks, a sympathetic reader, meeting this hypothesis in the late 1630s, rejected it as implausible, and sought a substitute model in the conical pendulum.[7]

As for Kepler, with the ellipse and area rule established to his own satisfaction, he could and did proceed to construct better planetary tables than anyone before him. More and more astronomers recognized that fact as the century progressed. It was the two rules in conjunction that were, in effect, getting verified.

To be sure, the astronomers did not accept either Kepler's hypothetical mechanisms or the areal rule derived from them. Boulliau, Wing and Streete adopted the ellipse but substituted other rules of motion for the areal rule.[8] They recognized, however, that they must closely approximate the predictions from that rule, as given, for instance, in Chap. 53 of the *Astronomia Nova*.[9]

In the *Philosophical Transactions* for 1670, Nicholas Mercator criticized Cassini for assuming the empty focus of the ellipse as an equant point; any rules proposed for the motion, he insisted, must closely approximate the areal rule.[10] Kepler's fitting of ellipse and areal rule to observations of Mars had become a standard against which astronomers tested their own alternative computational schemes. Flamsteed used the areal rule strictly from 1680. Cassini, seeking to retain an equant, replaced the ellipse by a new curve, now called the cassinoid.

Huygens, I believe, understood the logic of the situation. In a work of 1690, he spoke of "the elliptical movement that Kepler had divined (*deviné*), & verified (*verifié*) by observations."[11] The phrase "elliptical motion" conjoins ellipse and areal rule. I take "deviné" to mean "guessed", and "verifié" to mean "checked". The combination was getting checked, and was checking out.

1 On the History of the Phrase "Kepler's Laws"

Say, O Muse, who first spoke of Kepler's three rules as "Laws"? Leibniz, I believe it was, in his "*Tentamen de motuum coelestium causis*" of 1689. He referred to Kepler as "that incomparable man, whom the fates had watched over that he might be the first among mortals to publish the laws of the heavens, the truth of things, and the principles of the gods."[12] But, Leibniz added that Kepler

> "was not yet able to assign the causes to so many and so uniform truths, either because his mind was hampered by belief in intelligences or inexplicable sympathetic radiations, or because a more profound geometry and science of motions were not yet as advanced in his time as they are now."[13]

Leibniz attempted to account for Kepler's three rules using vortices and differential equations. As Bertoloni Meli had shown,[14] Leibniz, though explicitly denying it, wrote his essay only after making a close study of the first 40 pages of Newton's *Principia*. In his rivalry with Newton, he co-opted Kepler as an ally. By calling the rules "laws," he sought to enhance their status as results achieved independently of Newton.

Voltaire in his *Elements of the Philosophy of Newton* (1738) seemed to echo Leibniz. "Kepler," he said, "has merited the Name of Legislator in

Astronomy, notwithstanding his Philosophical Errors.... The extreme sagacity of Kepler discovered the Effect, of which the Genius of Newton has found out the Cause."[15] All three of Kepler's rules — Voltaire called them "laws" — are according to Voltaire effects which Kepler discovered but could not explain.[16]

For his account of Newtonian philosophy, Voltaire admitted to relying on commentators. 'sGravesande, proponent of Newtonian philosophy in Leiden, popularized the use of the term "law" for empirical results rendered general by induction. Voltaire attended his lectures in early 1737.[17]

On the assumption that the elliptical orbit was empirically established by Kepler, Voltaire had support from Henry Pemberton's *A View of Sir Isaac Newton's Philosophy* (1728). Pemberton said: "The strength of the centripetal force, in each place, was to be collected from the nature of the line, wherein the body moved. Now since each planet moves in an ellipsis, and the sun is placed in one focus, Sir Isaac Newton deduced from hence, that the strength of this power is reciprocally in the duplicate proportion of the distance from the sun."[18] Pemberton probably knew he was misreading Newton; he did so, in all likelihood, to make a difficult argument easier for his readers. The same ploy was used by Colin Maclaurin in his posthumously published *Account of Sir Isaac Newton's Philosophical Discoveries* (1748).[19]

The astronomer Lalande, in his *Abrégé d Astronomie* of 1774, appears to have been the first to list and number Kepler's three laws in the order in which they are customarily given today.[20]

The phrase, "Kepler's Laws", thus appears to have first arisen as a defensive maneuver in Leibniz's rivalry with Newton, and then to have been adopted as a popularizing ploy. For the ellipse and areal rule, the designation should be abandoned; it befogs our view of Newton's accomplishment.[21]

2 Newton's Knowledge of Kepler's Astronomy

In his letter to Halley of June 1686, Newton claimed that *he*, Newton, established the ellipticity of the planetary orbits. In the *Principia*, I remind you, the ellipticity of the orbits, as an empirical proposition, first emerged in Proposition 13 of Book III, where Newton asserted it as deducible from the inverse-square law of gravitation. Newton showed no sign of ever having read the *Astronomia Nova.* Whence his opinion that Kepler's ellipse was merely a *guess*?

In a long postscript to his letter of 20 June 1686, Newton, in rebuttal of Hooke's claim to originality in proposing the inverse-square law, cited a passage in Boulliau's *Astronomia Philolaica* of 1645.[22] Here Boulliau, in criticizing Kepler's celestial physics, reviewed all the main Keplerian assumptions: Keplerian inertia; the Keplerian motive virtue issuing from the Sun to push the planets round, its intensity falling off directly as distance; the aligned magnetic fibers in the planets. "Kepler's *figmenta*," Boulliau concluded, "are the offspring of an extremely ingenious mind, most clever in contriving causes of things where the true causes are hidden."[23]

Boulliau attacked Kepler particularly for choosing an inverse-distance variation for the solar virtue, rather than an inverse-square variation. Kepler had been the first to assign an inverse-square variation to light-intensity, on *a priori* grounds. Boulliau insisted that the same reasoning applies to the solar virtue. Newton in his postscript gave it as Boulliau's view that "all force respecting the Sun as its center & depending on matter must be reciprocally in a duplicate ratio of the distance from the center." Therefore, Hooke cannot claim the inverse-square law as his own invention.

To locate the statement he cites from Boulliau, Newton had to have read Boulliau's entire critique of Kepler's physics. This showed him that Kepler's mechanics was mistaken. Kepler, therefore, could not *know* the orbits to be elliptical. Boulliau, on the very next page, spoke of Kepler's choice of the ellipse as a "happy conjecture".[24] Perhaps that is an antecedent of Newton's statement about Kepler's having guessed the orbit to be elliptical.

It would not have occurred to Newton, I believe, to suppose that the ellipticity could be empirically established. An uncountable infinity of ovals neighbors on the correct ellipse; an infinity of them would fit the data.

3 Empirical Factors in the Emergence of the Main Argument of the *Principia*

What were the empirical factors and how important were they? This topic was dealt with penetratingly and in detail by William Harper and George Smith in their recent study, "Newton's New Way of Inquiry".[25] I attempt here only an *aperçu* of an historical moment. Between late December and late January 1684–1685, Newton and Flamsteed exchanged a few letters.

Twenty-nine years ago, when I first read Newton's letter of 30 December, his mood struck me as unusual. "I thank you heartily," he told Flamsteed, "for your kind information about those things I desired."[26] Again: "Your information about the Satellits of Jupiter gives me very much satisfaction."[27] And at the close of the letter: "A good new yeare to you."[28] For Newton, an exceptional quantity, wouldn't you say, of friendly good feeling?

Newton sent to the Royal Society the original version of his tract *De motu* in the preceding November. There he claimed to show that "the major planets orbit in ellipses having a focus at the centre of the Sun, and with their *radii vectores* describe areas proportional to the times, exactly as Kepler supposed."[29] He was assuming an exact inverse-square force towards the Sun. The first revision of the *De Motu* is tentatively assigned by Whiteside to December 1684. There, Newton articulated an idea nowhere expressed in the original tract:

> "... the whole space of the planetary heavens is either at rest... or uniformly moved in a straight line, and similarly the common centre of gravity of the planets... is either at rest or moved at the same time. In either case... the motions of the planets among themselves... take place in the same manner and their common centre of gravity is at rest with respect to the whole of space, and so it ought to be considered the immobile centre of the whole planetary system. Thence indeed the Copernican system is proved *a priori*. For if a common centre of gravity is computed for any position of the planets, this either lies in the body of the Sun or will always be very near it. By the displacement of the Sun from the centre of gravity it may happen that the centripetal force does not always tend to that immobile centre, and thence that the planets neither revolve exactly in ellipses nor revolve twice in the same orbit. Each time a planet revolves it traces a fresh orbit, as happens also with the motion of the Moon, and each orbit is dependent upon the combined motions of all the planets, not to mention their actions upon each other. Unless I am much mistaken, it would exceed the force of human wit to consider so many causes of motion at the same time, and to define the motions by exact laws which would allow of an easy calculation. Leaving aside these fine points, the simple orbit that is the mean between all vagaries will be the ellipse which I have discussed already."[30]

The implications are both wondrous and worrisome. If all celestial bodies mutually attract with the same kind of attraction, and no further causes are at work, then the natural philosopher can proceed to study the interactions mathematically, without regard to any aethereal hypotheses that might be proposed. Newton had previously invested time and thought in such hypotheses, but he was ever on the lookout for ways to transcend the merely conjectural. Assuming this attraction between all celestial bodies, he stated that the Copernican hypothesis can be established *a priori*. This result was important enough for him to make it the initial focus of the third book of the *Principia*, as originally written.[31] The attraction emerges out of a certain universalization: the Sun, Jupiter, Saturn attract; ergo all the celestial bodies attract. Newton seemed not to question the conclusion.

On the other hand, it is surely rather a hope than a sure thing, is it not? How to establish it as something of which we could have, if not logical certitude, at least what Locke and 'sGravesande called moral certitude — the sort of assurance on which our daily lives depend? And, assuming this general attraction, how to cope with the problem of the celestial motions, which has now become formidably intricate?

The principal topics in play in the Newton-Flamsteed exchange of December and January are these:

(i) Observations for computing the orbit of the comet of 1680–1681;
(ii) Application of Kepler's third or harmonic rule to Jupiter's satellites;
(iii) Application of this same rule to the planets, especially Saturn;
(iv) Satellite observations to determine the powers of attraction of Jupiter and Saturn;
(v) Perturbations of Saturn caused by Jupiter.

On the fifth topic, I shall say only this. Newton's suggestions for coping with these perturbations — namely, by referring Saturn's orbit to the center of gravity of Jupiter and the Sun, by introducing an oscillation in eccentricity and apsidal line, and by looking for special divagations in the years before and after conjunction with Jupiter — do not appear to have borne fruit for anyone. The remaining four topics I take up *seriatim*.

3.1 *Computing the orbit of the comet of 1680–1681*

In November 1680, a magnificent comet appeared in the eastern sky before sunrise, disappearing into the sunlight in early December. From 10 December to 9 March, what was taken to be either the same or another comet appeared in the western sky after sunset. Flamsteed, claiming the two apparitions to belong to a single comet, communicated to Newton his data and his notions about why the comet moved as it did. To Newton, it seemed more likely that the two apparitions belonged to different comets. If there was but one comet, that comet had swept nearly rectilinearly in toward the Sun, made a hairpin turn, then retreated in a direction nearly parallel to its incoming path. Such a comet, Newton thought, would be paradoxical:

> "The comets of 1665, 1677 & others which moved toward the Sun, ... had they twisted about the Sun & not proceeding on forward gone away behind him they would have been seen again coming from him. The many which have been seen advancing from the Sun... would have been

> seen in the former part of their course advancing towards him, had that former part been performed, not in the line of the latter part shooting on backwards towards the region beyond the Sun but twisting about him towards any hand."[32]

By the time he wrote his *De motu*, Newton had changed his mind. There, in Problem 4, assuming an inverse-square-law force of known magnitude towards the Sun, and positioning a body with given initial position and velocity, he showed that the conic-section orbit in which this body will move can be determined. In a Scholium, he added:

> "A bonus, indeed, of this problem, once it is solved, is that we are now allowed to define the orbits of comets, and thereby their periods of revolution, and then to ascertain from a comparison of their orbital magnitude, eccentricities, aphelia, inclinations to the ecliptic plane and their nodes whether the same comet returns with some frequency to us."[33]

Why this change of mind? It is a case, I say, of Newtonian-style universalization, as later articulated in the fourth rule of reasoning: propositions inferred by induction from phenomena are to be held accurately true, contrary hypotheses notwithstanding, till such time as other phenomena occur, by which they are rendered more accurate or liable to exceptions.[34] Two applications of just this rule occur in the objections Newton raised to Flamsteed's explanation of the comet's motion.

Flamsteed supposed both the comet and the Sun to be permanent magnets. Newton objected: the Sun is hot, magnets heated in the fire lose their magnetic powers; ergo, the Sun is not a magnet.[35] But, Flamsteed rejoined, the Sun's magnetism may be of a different kind. The Sun may not be hot, but only cause heat here, or perhaps only its outer layer is hot.[36] Newton, though aware that Aristotle and the medievals believed celestial bodies to be neither hot nor cold, will have none of this. Heat, like light, flows from the Sun, which he concluded must be vehemently hot.[37]

The other inductive generalization has to do with the relative strengths of the directive and attractive powers of magnets. Flamsteed supposed that the cometary magnet, after approaching the Sun with its attracted pole foremost, had been somehow turned about by the vortex so as to have its repelled pole facing the Sun, and so made its hairpin turn short of the Sun, and thus was repelled back in the direction from which it came. Newton objected:[38]

> "As for a great magnet exercising its directive vertue more strongly then its attractive on a small one, it holds in all cases I had opportunity to observe & till a contrary instance can be brought, I am inclined to believe it holds generally."

In the *De motu* of November 1684, Newton would have carried out another inductive generalization, by making comets along with planets subject to a single law of acceleration towards the Sun. The step here had to be conscious, bold. Alone among closely observed comets, the comet of 1680–1681, if indeed it was one comet, had "fetched a compass about the Sun" in a narrow hairpin curve. A common view, suggested by Hooke in 1678 and echoed by Halley, was that comets were bodies that had partially lost their gravitating principle.[39] Newton in 1681 knew not what to make of comets. In a sentence not sent to Flamsteed in 1681, he mentioned "a way of determining the line of a Comets motion (what ever that line be) almost to as great exactness as the orbits of the Planets are determined, provided due observations made very exactly be had."[40] He was wrong, and is still wrong about it in the *De motu*. The proposed method for finding the comet's tangential line of motion failed, as he found out by autumn 1685. The bonus will be obtained only much later. Determining cometary orbits was intended as the *pièce de résistance* of the *De Motu*, but in late 1684 it was only a hope.

3.2 *Kepler's harmonic rule as applied to Jupiter's satellites*

In December 1684, Newton requested from Flamsteed, along with data on the comet, data on Jupiter's and Saturn's satellites.

Flamsteed in his reply of 27 December is doubtful of the existence of the satellites of Saturn discovered by Cassini, that is, of satellites other than the Huygenian one. As for the Jovian satellites, he reported that their distances from Jupiter "are as exactly in sesquialter proportion to their periods as it is possible for our senses to determine."[41] Newton will repeat that statement in the first edition of the *Principia*. For the distance-determinations, Flamsteed employs a screw micrometer — an instrument only just coming into use. Only with the micrometer could the harmonic rule be demonstrated for Jupiter's satellites with impressive precision.[42] Flamsteed also assured Newton of the near-uniformity of the satellites' motions: their orbits are therefore very nearly concentric circles. But then, with a like precision, their accelerations towards Jupiter must be inversely as the squares of their distances from Jupiter. This conclusion is untroubled and definite.[43]

3.3 *The harmonic rule as applied to the circumsolar planets*

In his *De motu*, Newton said: "astronomers are agreed that the major planets follow the harmonic rule."[44] But in his letter of 30 December, he told Flamsteed that "The orbit of Saturn is defined by Kepler too little for the sesquialterate proportion."[45]

Newton's interest in Kepler's harmonic rule as applied to the planets goes back to the 1660s. He had learnt of it in Thomas Streete's *Astronomia Carolina* (1661). Streete followed Horrocks in claiming that, if you reduce solar parallax to 15 arcseconds, the rule comes out precisely true, and thus can be used to compute the mean solar distances of the planets from their periods, which are better known. This practice was Horrocks's innovation; it would make its way into the *Principia*, and with a small modification introduced by Newton, would be used ever after to derive mean solar distances from periods.[46]

In a manuscript first published by Rupert Hall,[47] and dating, apparently, from before 1669,[48] Newton found from Kepler's harmonic rule that the endeavors of the planets from the Sun vary inversely as the squares of their solar distances. He also compared the Moon's endeavor from the Earth with gravity at the Earth's surface; the result did not fit the inverse-square relation that I believe he would have been looking for.[49] In this manuscript he drew some consequences for astronomy,[50] but none from the inverse-square proportion; presumably because he saw not how to do it.

From 1669 or a little later, we have Newton's notes on the endpapers of Vincent Wing's *Astronomia Britannica* of 1669, first described by Whiteside in 1964.[51] Here Newton gave a purely empiric procedure for fitting orbital shapes and motions of planets to observations. He suggested referring certain of the lunar inequalities to the compression of the Earth's vortex by the solar vortex and he checked to see whether Wing's values for the planets' mean solar distances fit Kepler's harmonic rule.

Wing, differing from Streete, had disputed Horrocks's reduction in solar parallax, and did not use the harmonic rule to calculate solar distances. Newton found that in Saturn, Mars, and the Earth, where Wing approached the harmonic rule closely, his tables best agree with observations; and he suspected that if the mean distances of Mercury, Venus, and Jupiter were reduced to this rule, Wing's tables would better agree with observations. The harmonic rule by itself implies that Wing's planetary distances would lead to maximal errors of about 2 (minutes of arc) for Venus, 4.6 for Mercury, and 3 for Jupiter.

In December 1684, more was at stake for Newton. In the tract *De motu*, he wanted the centripetal forces to be accurately inverse-square. But in Kepler's *Rudolphine Tables*, Newton found that Kepler, the discoverer of the harmonic rule, had made Saturn's mean solar distance too small. See Table 1.

Table 1 Kepler's harmonic rule applied to his values for the periods and solar distances of the planets.

Planet	Sid. Period (Days)	Mean Distance in Rud. Tab. (K)	Mean Distance From Harmonic Rule (H)	100(K–H)/H
Saturn	10759.2	9.51004	9.53794	−0.293
Jupiter	4332.59	5.20000	5.20115	−0.022
Mars	686.981	1.52350	1.52369	−0.012
Earth	365.2566	1.00000	1.00000	
Venus	224.700	0.72413	0.72333	+0.111
Mercury	87.9691	0.38808	0.38710	+0.253

That worries him. Note that he did not worry about Kepler's exaggerated values of the mean distances for Mercury and Venus; long before, in the endpapers of Wing's book, he had noted that these orbits are expanded in appearance by refractions.

On 6 January, Flamsteed wrote this reply:[52]

> "As for the motion of Saturn I have found it about 27' slower in the Acronychal appearances since I came here, then Kepler's numbers, & Jupiters about 14 or 15' swifter..., the error in Jupiter is not always the same, by reason the place of his Aphelion is amisse in Kepler, nor is the fault in Saturn always the same..., yet the differences in both are regular & may be easily answered by a small alteration in the Numbers as is found in Saturn by our New Tables which Mr Halley made at my request & Instigation. I have corrected Jupiter myself so that he has of late years answered my calculus in all places of his orbit...."

Flamsteed thus suggested that once Kepler's numbers are corrected, all will be well. Newton's reply on 12 January embraced the thought:[53]

> "Your information about the error of Kepler's tables for Jupiter and Saturn has eased me of several scruples. I was apt to suspect there might be some cause or other unknown to me, which might disturb the sesquialtera proportion. For the influences of the Planets one upon another

> seemed not great enough tho I imagined Jupiter's influence greater then Your numbers determine it. It would add to my satisfaction if you would be pleased to let me know the long diameters of the orbits of Jupiter & Saturn assigned by yourself & Mr. Halley in your new tables, that I may see how the sesquiplicate proportion fills the heavens together with another small proportion which must be allowed for."

These three sentences tell us three things. First, given Flamsteed's confidence that Kepler's numbers can be corrected, Newton ceased to worry about unknown causes. Second, he knew how to compute the relative influence of Jupiter on Saturn and Saturn on Jupiter; more on that in a moment. This influence could conceivably be great enough to disturb the harmonic rule, though Newton believed it to be not so great. And third, Newton knew that the harmonic rule has to be slightly modified. The required modification is undoubtedly the one he worked out in Propositions 57–60 of Book I of the *Principia*, on the assumption that the Sun and each planet attract mutually.

Neither Newton nor Flamsteed suspected what Halley will later find: that the mean motions of Jupiter and Saturn, hence their periods, are changing over time. It follows that their mean solar distances are likewise changing. For the inner planets, mean solar distances assigned today agree to five places with the values given by Streete in 1661 and by Newton in 1687; in Jupiter the agreement holds only to three places; in Saturn only to two.[54]

With the consideration of mutual attraction between bodies, we are entering upon a new set of relations not broached in the original *De Motu*. We move from the one-body to the two-body problem, and the passage quoted refers in fact to the three-body problem. The development of the mathematical account in stages of increasing complexity and truthfulness to the world is what Bernard Cohen has called "the Newtonian style".[55] It is my conjecture that stage one was for Newton a gamble, entered on without a clear view of the implications of universalizing the centripetal forces so far assumed towards the Sun, Jupiter, Saturn, and the Earth; a stage at which Newton may conceivably have imagined stopping, with a worked-out orbit for a comet. But as he told Flamsteed in his letter of 12 January, "Now I am upon this subject I would gladly know the bottom of it before I publish my papers."[56]

Stage one will remain important as a first stage of an analysis. The basic ideas underlying Newton's *Principia* do not get their meaning and empirical warrant, one by one, from separate inductive generalizations; rather, they form a network.[57] The network shows its explanatory power when applied as a whole to observations. In December and January of 1684–1685, I think we are seeing a thus far successful explanatory scheme, though not the final one, clicking into place.

In the *Principia*, Newton solved the rhetorical problem of arguing for the harmonic rule as applied to the circumsolar planets by juxtaposing Boulliau's with Kepler's values for the mean distances; Boulliau's values are greater than Kepler's, and indeed exceed the values derived from the periods where Kepler's values fall short.

3.4 *Computing the absolute centripetal forces of Jupiter and Saturn*

The inverse-square law has to do with centripetal accelerations. The accelerations are inversely as the squares of the distances, independently of the nature of the body accelerated. Newton posed the question: what are the relative attractive powers of Jupiter and the Sun, their absolute centripetal forces as he will later call them? He could answer it at once, if both the Sun and Jupiter had satellites at the same distance; the ratio of the accelerations of the two satellites would be the ratio of the two attractive powers. Nature has not carried out this experiment for us. But there is no difficulty in carrying it out in thought. Using the inverse-square relation, we can compute the acceleration a satellite would have toward its primary at any stipulated distance. Thus, Newton was able to compute Jupiter's attractive power relative to the Sun's, and found it a little less than 1/1000th.

No one had ever done anything like this before. At some unknown moment, Newton must have been surprised and pleased. And the great bonus of this calculation was that he could establish the Copernican hypothesis *a priori*, as he put it. With Flamsteed's numbers, the center of gravity of Jupiter and the Sun comes outside the Sun by only 1/15th of the Sun's radius. From data on the Huygenian satellite of Saturn, Newton found Saturn's attractive power only half as great as Jupiter's; it is really less than a third. He found the Earth's attractive power 1/26th of Jupiter's, again too large owing to his taking solar parallax too large; it is really 1/318th. But the essential point was that the center of gravity of the system is never far from the Sun.[58]

Later (I would assume in the spring of 1685), Newton was able to show that the attractive powers of these different bodies were as their inertial masses.[59] The demonstration assumes that the third law of motion, action equaling reaction, holds.[60] Under an aethereal hypothesis, it might still be possible to account for the apparent attractions, though not easy.

When the universalization is extended to apply to every particle of matter in the universe, aethereal explanations are finally excluded, and Huygens and Leibniz will reject the result. It is not the logic of Newton's demonstration that is at fault; his rule of universalization takes him to surprising places. Propositions established by means of it are corrigible, liable to exceptions. But Newton would accept only an empirical refutation of his conclusions. In 1726, when Molyneux and Bradley found what they believed to be a nutation of the Earth's axis that could not — so they believed — be accounted for in Newton's system, anecdote has it that Newton responded merely by saying: "It may be so, there is no arguing against facts and experiments."[61]

The burden of my talk has been to say that Newton at the turn of the year 1685, having originally set out to solve some problems for the astronomers, had actually begun to uncover the staggeringly systematic and mathematical character of the System of the World.

Notes and References

1. *The Correspondence of Isaac Newton* (ed.) H.W. Turnbull (Cambridge, 1960), Vol. 2, p. 436.
2. Some authors, for example David Brewster (*The Martyrs of Science*, London, 1841, p. 263) and N.R. Hanson ("The logic of discovery," *Journal of Philosophy* **55** (1958): 1077 and *Patterns of Discovery*, Cambridge University Press, 1958, p. 107), claimed that Kepler's laws *implied* Newton's system.
3. Alexandre Koyré, *La révolution astronomique* (Paris, Hermann, 1961), p. 184.
4. Max Caspar, *Neue Astronomie* (Munich, Oldenbourg, 1929), Intro., pp. 46, 51.
5. See William H. Donahue, tr., *Kepler's New Astronomy*, p. 576; also Kepler's Gesammerie Werke, Vol. 3, p. 285.
6. D.T. Whiteside, "Keplerian eggs, laid and unlaid, 1600–1605," *Journal for the History of Astronomy* **v** (1974): 1–21, esp. 12–14.
   Kepler's mistaken claim was echoed by later explicators and defenders of Kepler's procedures: Robert Small, *An Account of the Astronomical Discoveries of Kepler* (London, 1804), pp. 304–305; C.S. Peirce in *Collected Papers* (eds.) Hartshorne and Weiss (Cambridge, Massachusetts, 1960), Vol. 2, pp. 53–54; N.R. Hanson, "The logic of discovery," *Journal of Philosophy* **55** (1958): 1077.

7. See my "On the origin of Horrocks's lunar theory," *Journal for the History of Astronomy* **18** (1987): 77–94, esp. 92–93. The orbit of the pendulum bob has to be displaced, as though by a wind, so that the orbit becomes eccentric. Horrocks assumed an inertia in the planet that includes a yen for the aphelion from which the planet started out. Horrocks's model influenced Robert Hooke's speculations about planetary motion.
8. See Curtis Wilson, "Predictive astronomy in the century after Kepler," in *Planetary Astronomy from the Renaissance to the Rise of Astrophysics. Part A: Tycho Brahe to Newton* (eds.) René Taton and Curtis Wilson (Cambridge University Press, 1989), pp. 172–185.
9. In Boulliau's *Astronomia Philolaica* of 1645, the page numbers 49 and 50 occur twice. As the printing of the book was nearing completion, Boulliau discovered that his procedure for determining the motion of Mars yielded heliocentric longitudes differing from Kepler's values by more than 2 arcminutes in the octants; the error in geocentric longitude could amount to three times as much. In the second set of pages 49–50, Boulliau simply gave Kepler's table of equations of center for Mars as a substitute for his own. (See Curtis Wilson, "From Kepler's laws, so-called, to universal gravitation: empirical factors," *Archive for History of Exact Sciences* **6** (1970): 115–116.) In 1657, he devised a new and more successful geometric procedure to replace the areal rule (*Ismaelis Bullialdi Astronomiae Philolaicae fundamenta clarius explicata & asserta, adversos clarissimi viri Sethi Wardi Oxoniensis professoris impugnationem* (Paris, 1657), on which see Wilson, *op. cit.*, pp. 119–120.)
10. "Some considerations of Mr. Nic. Mercator, concerning the geometrick and direct method of Signior Cassini for finding the apogees, excentricities, and anomalies of the planets, as that was printed in the *Journal des Scavans* of September 2, 1669...," *Philosophical Transactions* **V** (No. 57 for 25 March 1670): 1168–1175.
11. Huygens, *Oeuvres Complètes* **21**, 472.
12. From a translation by E.J. Collins, quoted by I. Bernard Cohen in "Newton and Keplerian inertia: an echo of Newton's controversy with Leibniz," in *Science, Medicine and Society in the Renaissance: Essays in Honor of Walter Pagel* (ed.) Allen G. Debus (New York, Science History Publications, 1972), Vol. 2, p. 205.
13. *Ibid.*

14. *See D. Bertoloni Meli*, "Public claims, private worries: Newton's *Principia* and Leibniz's theory of planetary motion," *Studies in History and Philosophy of Science* **22** (1991): 415–449 and *Equivalence and Priority: Newton versus Leibniz* (Oxford, Clarendon Press, 1993), pp. 95–142, Appendix 1.
15. I quote from the English translation by John Hanna, *The Elements of Sir Isaac Newton's Philosophy* (London, 1738), pp. 211–212. In the original, *Élémens de la philosophie de Neuton* (Amsterdam, 1738), the passage quoted is on p. 243.
16. *The Elements of Sir Isaac Newton's Philosophy*, p. 222; *Élémens de la philosophie de Newton*, p. 255.
17. Voltaire, *Correspondence*, (ed.) Theodore Besterman (Paris, Éditions Gallimard, 1963), 1704–1738, Vol. 1, pp. 826, 832, 841, 846, 866–867, 880, 963, 998, 1137, 1158, 1166. On 'sGravesande's thought, see his inaugural address at Leiden, *De matheseos in omnibus scientiis praecipue in physicis usu*, 1717, and his *Physices Elementa... sive Introductio ad philosophiam Newtoniam*, Leiden, 1720f.; see also Ernst Cassirer, *The Philosophy of the Enlightenment*, translated by Fritz C.A. Koelln and James P. Pettegrove (Princeton University Press, 1951), pp. 59–64.
18. Henry Pemberton, *A View of Sir Isaac Newton's Philosophy* (London, 1728), pp. 172–174.
19. Colin Maclaurin, *An Account of Sir Isaac Newton's Philosophical Discoveries, in Four Books* (London, 1748), pp. 47, 49, 266.
20. Joseph–Jérôme Le Français de Lalande, *Abrégé d Astronomie* (Paris, 1774), p. 201. I owe my information about this work to Professor Owen Gingerich of Harvard.
21. Newton himself, in the scholium he introduced in the third edition of the *Principia* following Proposition 4 of Book III, spoke of Kepler's harmonic rule as "legem planetarum a *Keplero* detectam."
22. Newton appeared to have consulted this work for the first time during the period of writing the *Principia.* In his letter to Flamsteed of 12 January 1684/1685 he said, "I have not at all minded Astronomy of some years till on this occasion which leaves me more to seek" (*Correspondence*, Vol. 2, p. 413). In the 1660s, Newton consulted the books on astronomy of Thomas Streete and Vincent Wing; in the 1670s, he was in communication with the astronomer Nicholas Mercator. His earliest reference to Boulliau appeared to be in his letter to Halley of 20 June 1686.
23. Ismaël Boulliau, *Astronomia Philolaica* (1645): 24.

24. *Ibid.*, p. 25.
25. William Harper and George Smith, "Newton's new way of inquiry," in *The Creation of Ideas in Physics* (ed.) Jerrett Leplin (Dordrecht, Boston, London, Kluwer Academic Publishers, 1995), pp. 113–166.
26. *Correspondence*, Vol. 2, p. 406.
27. *Ibid.*, p. 407.
28. *Ibid.*
29. *The Mathematical Papers of Isaac Newton*, Vol. 6, p. 49.
30. Quoted from the translation by A. Rupert Hall and Marie Boas Hall, *Unpublished Scientific Papers of Isaac Newton* (Cambridge, at the University Press, 1962), pp. 280–281.
31. See Florian Cajori (ed.) *Sir Isaac Newton's Mathematical Principles of Natural Philosophy and His System of the World* (University of California Press, 1947), pp. 549–563.
32. *Correspondence*, Vol. 2, pp. 364–365.
33. *The Mathematical Papers of Isaac Newton*, (ed.) D.T. Whiteside (Cambridge University Press, 1974), 1684–1691, Vol. 6, pp. 56–59.
34. *Isaac Newton's Philosophiae Naturalis Principia Mathematica. The Third Edition (1726) With Variant Readings* (eds.) Alexandre Koyré and I. Bernard Cohen (Harvard University Press, 1972), Vol. 2, p. 555.
35. *Correspondence*, Vol. 2, pp. 341–342.
36. *Ibid.*, p. 351.
37. *Ibid.*, pp. 359–360.
38. *Ibid.*, p. 360.
39. Robert Hooke, "Cometa," in *Early Science in Oxford* (ed.) R. T. Gunther VIII (Oxford, 1931), Vol. 8, p. 227; on Halley see *Correspondence*, Vol. 2, p. 351.
40. *Correspondence*, Vol. 2, p. 366.
41. *Ibid.*, p. 404.
42. See Florian Cajori (ed.) *Sir Isaac Newton's Mathematical Principles of Natural Philosophy and His System of the World* (University of California Press, 1947), p. 556.
43. Flamsteed mentioned in passing that the satellite periods must be taken when Jupiter is in his mean distance from the Sun. "For you well know that when Jupiter is on his Aphelium the satellites are made swifter, when Perihelius is slower" (*Correspondence*, Vol. 2, p. 404). *Did* Newton know this before? I doubt it. There is an exact analogy with the annual equation

of the Earth's Moon, and by and by Newton will account for both phenomena by solar perturbation.

44. *The Mathematical Papers of Isaac Newton* (ed.) D.T. Whiteside (Cambridge University Press, 1974), 1684-1691, Vol. 6, p. 41.
45. *Correspondence*, Vol. 2, p. 407.
46. Up to 1984, the periods and mean distances of the inner planets from Mercury to Mars were treated as basically fixed, and variations treated as perturbations; while for the outer planets only osculating elements (including mean distances) were assigned. See, for example, *Explanatory Supplement to the Astronomical Ephemeris and the American Ephemeris and Nautical Almanac* (London, Her Majesty's Stationery Office, 1961), pp. 111–115. More recently, only osculating elements are computed for all planets.
47. A. Rupert Hall, "Newton on the calculation of central forces," *Annals of Science* **13** (1957): 62–71.
48. *Correspondence*, Vol. 1, p. 301, n. 1.
49. *Correspondence*, Vol. 2, p. 436. The inverse-square relation would make the force countervailing the Moon's endeavor to recede analogous to the similarly functioning force towards the Sun. Also, it was a consequence, as Newton would later assert (*Correspondence*, Vol. 2, p. 440), of an aethereal hypothesis for gravity which he had first formulated in a student notebook (see J.E. McGuire and Martin Tamny, *Certain Philosophical Questions: Newton's Trinity Notebook* (Cambridge University Press, 1983), pp. 362–365, and later elaborated in his "Hypothesis explaining the properties of light," sent to the Royal Society in December 1675 (*Correspondence*, Vol. 1, 365–366).
50. As Hall remarked, "Newton's interest was in the calculation of centrifugal accelerations for the sake of their application to astronomical problems" (op. cit., n. 48, p. 63). There were two such applications.

    In the first, assuming the Earth's rotation, Newton calculated the endeavor to recede of a body at the Earth's equator, and found it only 1/350th of the acceleration of gravity, too small to cause bodies to "leap into the air" and "fly from the Earth" (Newton's phrases). Galileo's earlier attempt to refute this Ptolemaic objection to the Earth's rotation was much less satisfactory (see Galileo Galilei, *Dialogue Concerning the Two Chief World Systems — Ptolemaic & Copernican*, translated by Stillman Drake (University of California Press, 1962), 198ff.

The second practical application was hypothetical: *if* the reason the Moon always turns the same face towards the Earth is the Moon's endeavor from the Earth's center, then this endeavor must be greater than the Moon's endeavor to recede from the Sun. The inequality will imply that the horizontal solar parallax, a disputed constant at the time, and crucial in constructing solar theory, is greater than 19 arcseconds. But Newton's inference was wrong: The Moon's endeavor from the Earth is *not* greater than its endeavor from the Sun, and the solar parallax is *not* greater than 19 arcseconds. The dynamical idea here is muddled. In a letter of 1673 to Oldenburg for Huygens, Newton seemed to be taking a step toward disentangling himself from the muddle (*Correspondence*, Vol. 1, p. 290). Eventually, his line of thought will lead him to the discovery of the Moon's physical libration in longitude. The adequate quantitative treatment of this topic, however, will have to wait upon Lagrange, for it requires rotational dynamics, which Newton never realized he lacked (see Curtis Wilson, "The work of Lagrange in celestial mechanics," *Planetary Astronomy From the Renaissance to the Rise of Astrophysics*, Part B: *The Eighteenth and Nineteenth Centuries* (eds.) René Taton and Curtis Wilson (Cambridge University Press, 1995), pp. 109–112.

51. Derek T. Whiteside, "Newton's early thoughts on planetary motion: a fresh look," *British Journal for the History of Science* **2**(6), (December 1964): 117–129.
52. *Correspondence*, Vol. 2, p. 408.
53. *Ibid.*, p. 413.
54. See *Planetary and Lunar Coordinates For the Years 1984–2000* (London, Her Majesty's Stationery Office; Washington, U. S. Government Printing Office, 1983), pp. 312–313.
55. I. Bernard Cohen, *The Newtonian Revolution* (Cambridge University Press, 1980), p. 62 ff.
56. *Correspondence*, Vol. 2, p. 413.
57. On the "network" character of Newton's laws, see Brian Ellis, "The origin and nature of Newton's laws of motion," in *Beyond the Edge of Certainty* (ed.) Robert G. Colodny (Prentice-Hall, 1965), pp. 29–68; and Ernan McMullin, "The significance of Newton's *Principia* for empiricism," in *Religion, Science, and Worldview: Essays in Honor of Richard S. Westfall* (eds.) Margaret J. Osler and Paul Lawrence Farber (Cambridge University Press, 1985), pp. 33–59.

58. For an explanation of how the Copernican system can be demonstrated "*a priori*" without presupposing universal gravitation, and how the demonstration nevertheless led Newton to the idea of universal gravitation, see the article cited in n. 25, pp. 134–139. The explanation is due to George Smith.
59. In Proposition 69 of Book I and Proposition 7 of Book III.
60. For a discussion of the argument, I refer you to Howard Stein's "From the phenomena of motions to the forces of nature: hypothesis or deduction?" in *PSA 1990: Proceedings of the 1990 Biennial Meeting of the Philosophy of Science Association* (eds.) Arthur Fine, Micky Forbes, and Linda Wessels (Philosophy of Science Association, East Lansing, Michigan, 1991), Vol. 2, pp. 209–222; and to Dana Densmore's commentary on Prop. 7 of Bk. III, Newton's Principia: The Central Argument (Green Lion Press, 1995), p. 353 ff.
61. Sir David Brewster, *Memoirs of the Life, Writings, and Discoveries of Sir Isaac Newton* (Johnson Reprint Corporation, 1965), Vol. 2, p. 407.